Lewis Creek Lost and Found

MIDDLEBURY BICENTENNIAL SERIES
IN ENVIRONMENTAL STUDIES

Christopher McGrory Klyza and Stephen C. Trombulak,
The Story of Vermont: A Natural and Cultural History

Elizabeth H. Thompson and Eric R. Sorenson,
Wetland, Woodland, Wildland: A Guide to the Natural Communities of Vermont

John Elder, editor,
The Return of the Wolf: Reflections on the Future of Wolves in the Northeast

Kevin Dann,
Lewis Creek Lost and Found

Lewis Creek Lost and Found

KEVIN DANN

May 2013

For Henrik –

Many thanks for helping us to launch ENIGMA!

Best wishes –

Kevin

MIDDLEBURY COLLEGE PRESS

Published by University Press of New England • Hanover and London

Middlebury College Press

Published by University Press of New England, Hanover, NH 03755

Printed in the United States of America 5 4 3 2 1

Decorative initial capitals are by Rachel Robinson, from Rowland E. Robinson, *Hunting Without a Gun* (New York, Forest and Stream Publishing Co., 1905).

Library of Congress Cataloging-in-Publication Data

Dann, Kevin T., 1956–

Lewis Creek lost and found / Kevin Dann.

p. cm. — (Middlebury bicentennial series in environmental studies)

Includes bibliographical references (p.) and index.

ISBN 1–58465–071–0 (cloth : alk. paper) — ISBN 1–58465–072–9 (paper : alk. paper)

1. Natural history—Vermont—Lewis Creek. 2. Bioregionalism—Vermont—Lewis Creek.

3. Lewis Creek (Vt.)—History.

I. Title. II. Series.

QH105.V7 D36 2001

508.743'17—dc21 00–010581

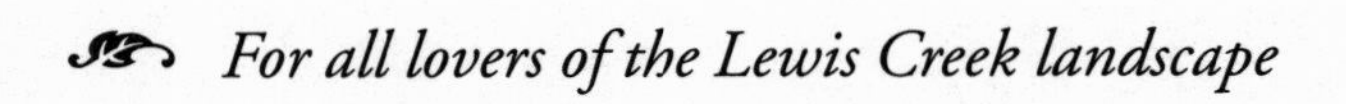
For all lovers of the Lewis Creek landscape

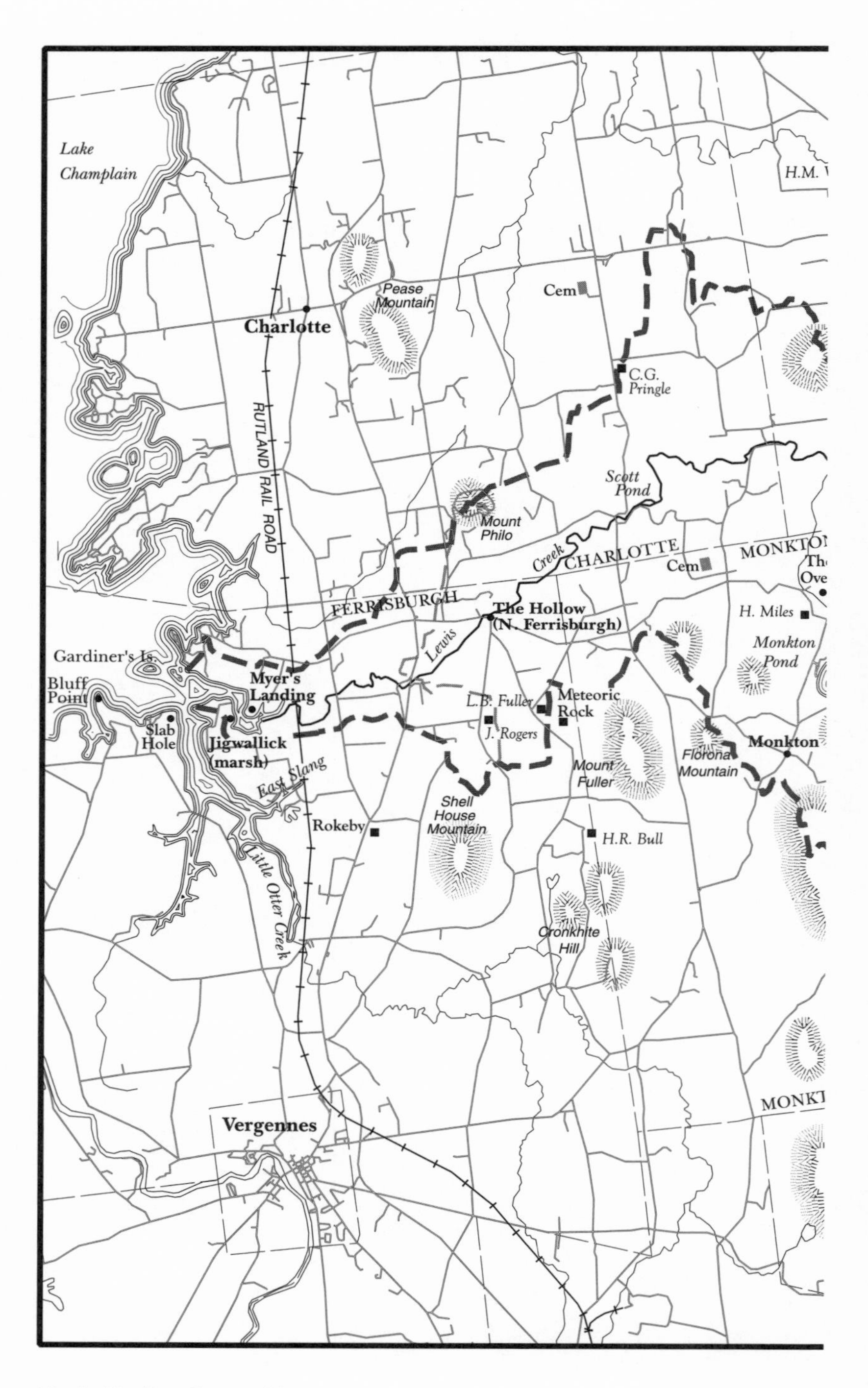

Map by Northern Cartographic

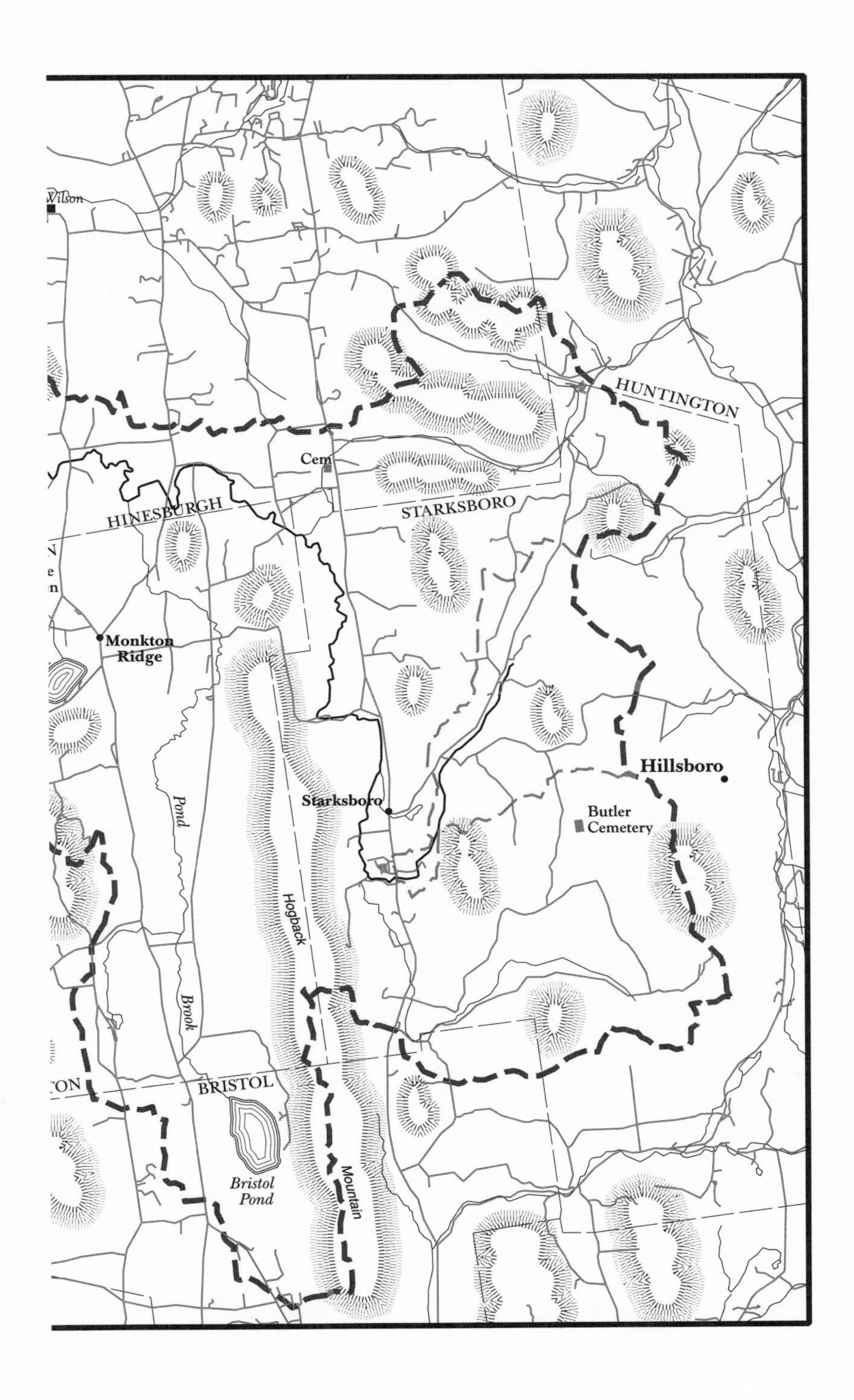

Wilson
HUNTINGTON
Cem
HINESBURGH
STARKSBORO
Monkton
Ridge
Hillsboro
Starksboro
Butler
Cemetery
Pond
Hogback
Brook
BRISTOL
Mountain
Bristol
Pond

If one stays beneath the star he was born under,
watching and waiting, it may, at last, prove a lucky one.
—Rowland Evans Robinson, "In Search of Nothing"

Contents

Illustrations

FIGURES

MAPS

Foreword

Stephen C. Trombulak

Lewis Creek Lost and Found is part of the Middlebury College Bicentennial Series in Environmental Studies, cosponsored by Middlebury College and the University Press of New England. Each of the books in this series adopts a bioregional approach to environmental topics. Such an approach emphasizes the continuity between natural history and human history, and often seeks to illuminate such connections by focusing closely on the history and characteristics of particular landscapes. The inclusiveness of bioregionalism is a natural outgrowth of the complex environmental history of New England and the Adirondack Mountains. Wild nature and human cultures are intricately interwoven in this region, which although long settled by humans has recently experienced a dramatic resurgence of forests and wildlife. The editors of the Middlebury Series believe that a healthy irony can enter into environmental discourse through study of this region's turbulent history and surprising present, and can perhaps illuminate a possible pathway for environmental recovery in other regions of the world.

Lewis Creek Lost and Found fills a special place in this series. Through the lens of historical narrative, it focuses on one particular watershed in western Vermont. At one level, this book offers up the biographies of three men—Cyrus Pringle, Rowland Robinson, and John Perry, individuals whose lives and searches in the Lewis Creek watershed, over almost a hundred years in the 1800s and early 1900s, exemplify many of the prevailing traditions about the human–nature relationships of the time. These men, and the other people whose lives were intertwined with theirs, are examples of local heroes, stewards of one specific place in time.

At a deeper level, however, *Lewis Creek Lost and Found* is the story of what it takes to inhabit, as opposed to merely live in, a landscape. Kevin Dann explores the spoken languages—both Abenaki and the local English vernacular—that signify the relationships people have with the nature of a landscape. It is also a story of the diverse cultures that have shaped both the natural history and human history of the region: Abenaki, French-Canadian, African-American, English, Irish, and Scottish. It is a sympathetic treatment of the challenges all people faced in trying to maintain their identities while working to create lives of meaning in a place that was quite hard and often unforgiving. Yet Dann's story also connects the lives of the people living in the Lewis Creek watershed to broader landscapes: neighboring watersheds, nested watersheds, Vermont, New England, and ultimately the entire continent. His language reinforces the truth that cultural history and natural history are intertwined and interpretable through shared metaphors: people and places existing at the margins; native and exotic residents and species; evidence of past lives writ on rocks and headstones.

Lewis Creek Lost and Found creates a narrative that is neither romantic nor idealized but is as rough and diverse as the land and the people themselves. As we seek to understand how cultures can come to reinhabit a landscape, it is the stories revealed by its earlier residents that help show us the way.

Acknowledgments

My role as writer of this book has been much like that of a river: I've acted as a conduit for the anecdotes and thoughts of the countless people who in a sense are the tributary streams of this story. All have contributed differently, though the final responsibility for the delivered volume is entirely my own. My thanks to Special Collections at Bailey-Howe Library and the Pringle Herbarium, University of Vermont, and Rokeby Museum for permission to use manuscript material and to reproduce illustrations in their collections. My heartfelt thanks to the following people: Thomas Altherr, David Barrington, Tom Bassett, Elinor Benning, David Blow, Michael Canoso, Charlie Cogbill, Mr. and Mrs. Norton Davis, Gordon Day, Evelyn Dike, Pru Doherty, Connie Gallagher, Kevin Graffignino, Bertha Hanson, Linda Henzel, Elizabeth Hinsdale, Susan Hopp, Charles Johnson, Bob Lagor, Dean Leary, Steve Loring, John McCreary, Everett Marshall, Ursula Marvin, John Moody, Dana and Christine Morgan, Happy Patrick, Roberta Poland, Marylee Rose, Greg Sanford, Ed Steele, Richard Sweterlitsch, Peter Thomas, Liz Thompson, Jane Williamson, Rod Wentworth, Ian Worley, Peter Zika. Nancy Gallagher and Bill Howland helped me immensely with their criticism and support. I am especially grateful to John Elder for taking an interest in the book, and to Steve Trombulak for so conscientiously overseeing its preparation.

K.D

Lewis Creek Lost and Found

Champlain Valley in vicinity of Lewis Creek, ca. 1860, by R. E. Robinson. Courtesy Rokeby Museum, Ferrisburg, Vermont.

Introduction

We have an unknown distance yet to run, an unknown river to explore. What falls there are, we know not; what rocks beset the channel, we know not; what walls rise over the river, we know not. Ah well! We may conjecture many things.
—John Wesley Powell

HIS IS NO river journey in the tradition of John Wesley Powell and the Colorado, nor John Stanley and the Nile, nor Huck Finn and the Mississippi, not even Thoreau and the Merrimack. Lewis Creek is a middling river, anonymous and unseen to all but a few fishermen, canoeists, and the neighbors who live along or not far from its banks. It boasts no spectacular falls, no epics of adventure and discovery. Though a part of one of North America's largest watersheds—the St. Lawrence—it is itself one of Lake Champlain's lesser tributaries, draining an area of only about 100 square miles. No warships have been built on it like on Otter Creek; no historic figures are associated with it like the Allen brothers and the Winooski. Its pleasures are small, intimate ones, the very sort that can be found along every river of any size in the New England region.

I began haunting the banks of Lewis Creek twenty years ago, after I first moved to Vermont. Though my family and I settled in the La Platte River watershed just to the north, I always found myself drawn to Lewis Creek as an avenue of exploration of Vermont's Champlain Valley landscape. While a canoe trip down the La Platte frequently brought the sights and sounds of the present into view, when canoeing or walking along Lewis Creek, one was in the presence of what seemed largely a nineteenth-century landscape. Like so many people who have moved to northern New England in the past few decades, settling in Vermont was for me an attempt to go back in time, to flee the placelessness of the modern world for a much mythologized past. I had grown up in northern New Jersey, at the edge of metropolitan New York's expanding megalopolis, and during my childhood the last few farms that surrounded our home were transformed into commercial and industrial parks. The brook that ran by our house was for the children of our neighborhood a route for exploration and discovery, and we used the brook as a way to escape time. We usually did this by going upstream, for the further one progressed downstream, the more denuded and despoiled were the banks, the more polluted was the water, and the more illusory was the sense that we were the brook's original explorers. Upstream, we could indulge our fantasy, as roads and houses gave way to fields and woodlots, and cellar holes and other ruins that only added to our mission of discovery.

Like that little brook, Lewis Creek and all of New England's streams both large and small offer us, as they did first the region's aboriginal inhabitants and later the European colonists, avenues to explore both space and time. In this well-watered land, we all live within a short walk to some channel draining inexorably toward the sea, and whether we follow it upstream or down, there is no more inviting guide to get to know one's place than stream, brook, and river. When Henry David Thoreau set off with his brother John in August 1839 up the Concord River, bound for the Merrimack and New Hampshire's "wilderness"—relative to settled Concord—he felt that the river offered more than just a convenient physical route for travel. "The river was the only key which could unlock [New Hampshire's] maze, presenting its hills and valleys, its lakes and streams, in their natural order and position." Their *natural* order; for Thoreau, and for us, there is something inevitable and intuitive about rivers—especially the dendritic patterns of New England's rivers—as a device for the acquisition and classification of new knowledge. Their geometry mirrors our own inner geometry, be it our branching nervous or circulatory system or the ramifying rays of our thoughts.

As intuitive as rivers might seem to be for getting to know a place, there is historically a surprising paucity of New England regional literature that chooses fluvial routes of discovery and exploration. In the late nineteenth century—the very same era that saw America's principal arid region, the

Southwest, mapped and described centrifugally from the Colorado River and other waterways—river-dissected New England entered its heyday of local description, as urban New Englanders traveled by rail to see the region's sublime attractions of lakes and mountains. The railways' promotional literature directed fishermen to trout streams, but these riverine destinations were the sites of episodic Arcadian refuge rather than wide highways for the discovery of inner, as well as outer, landscapes. The 1870s and 1880s saw the flourishing of local history in New England, but political boundaries—towns and counties—formed the framework for these important agents of collective memory.

In those same decades, sportsmen, scientists, and conservationists were just beginning to widen their attention from what lay between the banks, to the units of landscape—the watershed—defined by the rivers. Borrowed by midcentury from the German *Wasserscheide,* "separation of the waters," scientific usage of the term "watershed" began in nineteenth-century Great Britain, where naturalists like Charles Darwin and Charles Lyell used it to describe the divide itself. In American usage—at first largely confined to physical geographers—"watershed" moved away from its sense of referring to the division of waters upon the landscape, toward the sense of the land unit that *collected* surface water. Spatially the watershed encompasses all the land and water area between the divides, that is, the imaginary boundary lines dividing the flow of rainfall.

Watersheds collect stories along with precipitation. Their soils hold the histories of generations of past inhabitants, and release them slowly to the present. When Thoreau oared his dory up the Concord and Merrimack, his thoughts spilled over the banks and into the enveloping landscape, touching as frequently upon the heavens above as the earth below. When he abandoned the boat at the Merrimack's last lock, to press on afoot toward the headwaters, and imagined himself buried up to his chin in a swamp, "scenting the wild honeysuckle and bilberry blows," Thoreau discovered one more portal toward transcendence of the physical world. Ranging abroad in search of fuel for his campfire, hatchet in hand, he always came back with more than firewood. As river guides, he as frequently employed Hesiod, the New Testament, and the Bhagvad Gita as he drew upon Massachusetts and New Hampshire colonial historians Daniel Gookin and Jeremy Belknap.

Nearly a century after Thoreau's account of his river pilgrimage was published, the major rivers of New England, and all of America, were chronicled by dozens of the nation's finest writers, in the "Rivers of America" series, published between 1937 and 1949. The series demonstrated a pronounced northeastern chauvinism, as it included, along with the Hudson, Delaware, and Connecticut, lesser rivers like the Kennebec, Charles, Raritan, Passaic, St. Johns, Housatonic, Mohawk, and even the Winooski River. All of the books were primarily histories, and so the long-settled eastern seaboard, having cradled the lion's share of America's colonial past, received greater attention. Though as variegated in style and scope as their subject streams, the Rivers of America books shared a fascination for the heroic and romantic. They invariably began with sketches of the native peoples of the region, moved through tales of European colonial struggle and conquest, both against the Indians and each other, and mixed broad economic and social description with closely drawn portraits of individuals. The books were no different than their times in their relative blindness to the stories of women, ethnic and racial minorities, and other "people without history," but they continue to serve as effective entry points into regional American history and natural history.

The Winooski River's inclusion suggests that if a sufficiently mythical historical figure were associated with a river, that river might merit a place in the Rivers of America canon. Ralph Nading Hill's 1949 book, *The Winooski: Heartway of Vermont,* captured the tales of such idiosyncratic characters as miracle mathematical calculator Zerah Colburn, Methodist circuit preacher Lorenzo "Crazy" Dow, electromagnetic machine inventor Tom Davenport, and the von Trapp Family Singers, but the unmistakable hero of the book was Revolutionary War hero, land speculator, and frontier philosopher Ethan Allen. The Winooski provided a convenient backdrop against which Hill narrated the exciting exploits of Ethan Allen and the Green Mountain Boys. Breezy biography makes the Winooski run in Hill's book, as it does in a majority of the other works in the Rivers of America series. Only occasionally, as in Marjorie Stoneman Douglas's monumental work, *The Everglades,* did the river and its surrounding landscape take on the qualities of a heroic character. And none of the heroes portrayed in the Rivers of America books were conservationists; in Ralph Hill's *Winooski,* Vermont Governor George Aiken and senators Ernest Gibson, Sr., and Warren Austin come closest to that role, in their 1930s fight against the Waterbury dam, the Roosevelt administration, and the Green Mountain Power Corporation.

Today, a half century after the publication of the Rivers of America series, it is hard to imagine that any regional description hung upon the framework of a river would not feature both the landscape and its students—professional natural scientists—and defenders—grass-roots environmental activists—as principal characters. At century's end, no New England river of any size lacks dedicated stewards working to preserve and restore the ecological health of the waters and their watersheds. But New England's contemporary riverkeepers, though possessing enormous amounts of biological, geological, and hydrological information unavailable to earlier generations, frequently are unaware of the efforts of their predecessor students and stewards. From the Connecticut shoreline to the upper reaches of the Kennebec, from Cape Cod to Lake Champlain, New Englanders know Concord's Henry Thoreau, while they seldom know the names and deeds of past lovers of their own local watershed.

If we are to cultivate bioregionalism as a means of attaching ourselves to place, to "reinhabiting" the landscape, then we need to recover the stories of "local heros," the lesser known individuals who have both physically stewarded a place, and "metaphysically" stewarded a place through their cultivation of thought and memory. We perhaps stand in as great a danger of losing these histories as we do of losing the inviting bioregional landscapes that serve as the primary vehicle of our loyalty. In this book, I attempt to follow Lewis Creek in the direction of some of those local heroes, whose discoveries provide a bridge for me to make my own discoveries about place and the past. The individuals who have loomed largest are three men whose words remain as a record of their meanderings, as surely as the oxbows and terraces tell of Lewis Creek's past course: Rowland Evans Robinson (1833–1900), an author and illustrator whose artistry sprang from a deep love of story and place; Cyrus Guernsey Pringle (1838–1911), the "prince of plant collectors," who knew better than anyone the plants of the Lewis Creek watershed from Gardiner's Island to Hogback Mountain; and John Bulkley Perry (1825–1872), whose twin call to God and geology makes his writings a wonderful window into the late-nineteenth-century imperative to reconcile religion and science. All three lived during a time when science was transforming American life, and all had vocations that were impacted by, as well as spoke to, the changes that science and technology were effecting. More than in any other era, the roots of Lewis Creek's present seem to lie in that historical moment between 1860 and 1900. Though a century away, it seems close and contemporary. The same sights and sounds elate us, and the same tragedies befall us.

Whatever we find, it is possible to lose, and in this book loss is as much a part of the story as is discovery, the traditional motif of river journeys. Cyrus Pringle, Rowland Robinson, and John Bulkley Perry—and their contemporary late-nineteenth-century New England naturalists—were all engaged in processes of discovery and exploration in New England, in an era that we typically think of as witnessing exploration and discovery only in the newly conquered regions of the American West. Pringle, Robinson, and Perry are all discoverers, "finders": Pringle finds plants; Robinson finds lost folkways, habits of speech, and even lost landmarks in both space (his discovery of the grey pine) and time (his chronicling of the spearing of the last salmon in Lewis Creek); Perry peers deep into the past and finds fossils and structural relations that explain the regional geological story.

New England, and particularly Vermont, is a region where myths about the past are particularly tenacious, perhaps because it holds such an important place in our conceptions about our national culture and character. Hopefully, the stories told here—whether of a Quaker plant collector's love affair with a hired man, the Vermont eugenics movement's tragic marriage of flawed science to class and ethnic prejudice, or the ironic twists in the naming of a pond—can suggest new routes of discovery and exploration for any watershed, large or small. Thoreau believed that "it is easier to discover such a new world as Columbus did, than to go within one fold of this which we appear to know so well." Let us look upon our region's lesser rivers as new worlds, where within their folds lie ample discoveries for all of us.

CHAPTER 1

Footfall

NE NEEDS TO find some footing before setting off up the Creek, some place that would best make a beginning for this riverine journey. The headwaters would seem the natural place to start, the mouth the place to finish. But in reaching for some unnamed outer branch of the living tree that is the Lewis Creek watershed, the waters become quicksilver, magically shifting their position. In geologic terms, they are as elusive and impermanent as mercury, for what is a headwater today lay over the divide in the La Platte or the Little Otter Creek watershed yesterday, or may tomorrow. Then there is the question of *which* headwater—the little rivulet coming off the west flank of Lincoln Hill in Hinesburg, or the bubbling spring that issues from the east slope of Shaker Mountain in Starksboro, or the muddy stream that lies at the foot of Hogback Mountain in Bristol, snaking its way through cow pasture and beaver-flooded cedar swamp to Bristol Pond? Headwater streams are too ephemeral, and too numerous, to mark this beginning.

Just as a mind wants some solid island of knowledge and belief to stand on when confronting something vast, this river journey calls for some place immutable from which to gain a larger view. It should be some island that stands outside, yet is part and parcel of that view. There is an island a few hundred yards from the mouth of Lewis Creek that today goes by the name Gardiner's Island, and it seems a good place to begin. For at least the last few thousand years, every grain of sand, every uprooted elm, every drop of water that has ever made its way down the Creek to the mouth has spilled out in front of Gardiner's Island. Though rooted in the Paleozoic bedrock that cradles Lake Champlain, the island itself is like a big limestone cobble spit forth by the Creek. One can imagine it tumbling downstream during some long-past spring flood to come to its resting place in Hawkins Bay. From it one can look east into the wild rice and cattail marshes that define the mouth of the Creek. To the west looms Split Rock Mountain, which brings the ancient crystalline Adirondack rock right down to the edge of the lake.

But underfoot one's boots don't scrape craton—continental heartland—as they would across the lake. On Gardiner's Island they scuff limestone, a calcium carbonate extravaganza in the form of the pulverized shells of benthonic invertebrates. Over one hundred feet and as many million years deep, the stage of earth history represented here geologists today call the Crown Point formation, named for the type locality further south at the embattled peninsula of that name. In the exposed faces of the massive bluish-gray limestone you can see the remains of a few unpulverized animals: brachiopods, which fed on the rich rain of detritus that emptied into an ancient sea; nautiloids, with their uncanny mathematical symmetry; and the colonial "moss animals" known as bryozoans, looking like gumdrops, lace, twigs—anything but animals. Most visible of all though, is a larger denizen of those Paleozoic marine waters—*Maclurites magnus.*

Early settlers in Vermont regarded the spiral coils of this fossil gastropod as petrified snakes, but by the mid nineteenth century, paleontologists knew well that the conspicuous coils belonged to a big marine snail that they originally named *Maclurea magna.* It was especially characteristic of the rock formation that was then known as the "Chazy Limestone," which outcropped in the Champlain Valley from Chazy on the western shore of Lake Champlain, through Isle La Motte, South Hero, and some of the eastern lakeshore's most spectacular headlands, like Bluff Point (the southern headland of Hawkins Bay) and Grosse Point a little farther south. For nineteenth-century geologists seeking to cipher the epic tale told by the Paleozoic rocks of the Champlain Valley, the lake provided a fine avenue of exploration. Both high cliffs and shelving shores gave much greater glimpses of exposed bedrock than could be found in miles and miles of tramping away from the lakeshore.

Amateur paleontologist John Bulkley Perry knew the Chazy rocks and *Maclurea magna* well from Swanton, Vermont, where he was the Congregational minister in the 1850s. There, the big spirals marked the face of the town's most prominent buildings, for the Chazy limestone was sawn at a mill at Swanton Falls on the Mississquoi River. As a boy growing up in Burlington, he used to see *Maclurea* staring out from curbstones, steps, and lintels of houses all over town. But Gardiner's Island was perhaps the perfect place in which to contemplate the fossil snail and its ancient environment.

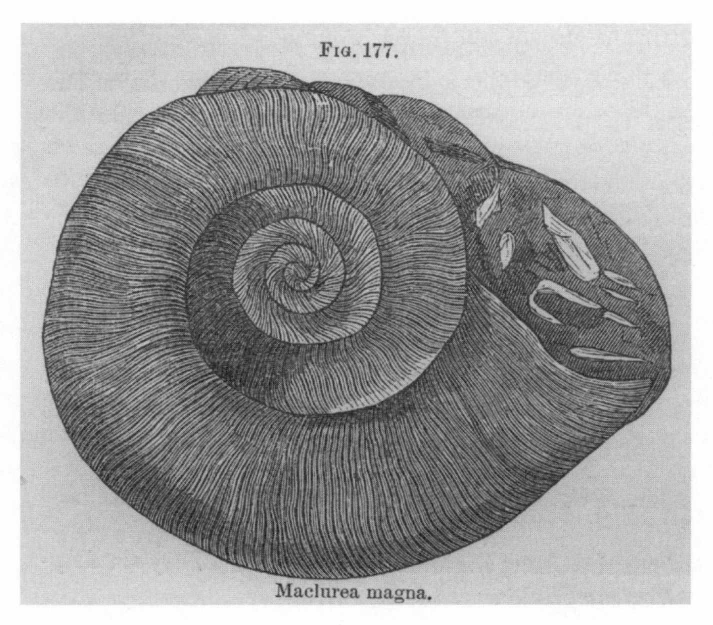

Maclurea magna, the nineteenth-century taxonomic term for the marine snail now known as *Maclurites magnus.* From Edward Hitchcock et al., *Report on the Geology of Vermont—Descriptive, Theoretical, Economical, and Scenographical,* 2 vols. (Claremont, N.H.: Claremont Manufacturing Co., 1861).

Sitting on the open ledge on the island's western shore, Perry could look straight across the lake to Split Rock Mountain, with its ancient "azoic" Adirondack rock in plain view. Hundreds of millions of years ago, the Adirondack massif was surrounded by a shallow sea that provided home for *Maclurea* and the other animals now frozen in the island's bedrock. Surrounded by waves beating against Gardiner's Island's shore, Perry found it easy to imagine the ancient sea. He also found that the expansive view from the island, toward headlands where he had collected fossils of older and younger strata, allowed him to picture a long sequence of shifting paleoenvironments to fit the geologic evidence. He could even see from his island perch the summits of Mount Philo, Mount Fuller, and Shellhouse Mountain—the "Red Sandrock" hills that brought the base of the Paleozoic into clearer view.

Perry's Gardiner's Island vista up and down the lake and into Paleozoic time, however, did not afford him the most critical view of all—into the very beginnings of life on earth. The waters Perry looked out upon from Gardiner's Island were the deepest in Lake Champlain; the steep palisades of Split Rock Mountain continued to plunge for hundreds of feet beneath the surface. They were as impenetrable as the mystery that lay beneath the surface of the bitter scientific debate then raging over the order of the earliest Paleozoic strata—just when did the earth first become a *living* planet, and how did that life arise? Perry felt the question more deeply than almost any of his fellow geologists.

Perry's contemporary and the authority on the Chazy rocks, Ebenezer Emmons, had described *Maclurea magna* as "sinistrorsal, discoidal, depressed turbinate, breadth more than twice as great as the height; spire flat, a slightly depressed line at the sutures; whorls about six, gradually increasing from the apex, ventricose, flattened above, obtusely angular on the outer edge; surface marked by fine striae, which, upon close examination, are found to be produced by the outer edge of the laminae, striae undulating, bending backwards from the sutures, and forwards in passing over the ridge of the shell; aperture obtusely trigonal, depressed above, slightly expanded beyond the dimensions of the whorl just behind it; axis hollow; umbilicus broad and deep, extending to the top of the spine." This thick description obscures the fact that *Maclurites* encompasses worlds in its geometry. Like so many seashells, *Maclurites* spirals so that the progressively larger domains of the creature wrap around its previous domains in a regular manner. As the living *Maclurites* grew in its Paleozoic home, its coils increased in width to produce what is called a logarithmic or equiangular spiral, or what James Bernoulli aptly deemed the *spira mirabilis*. The cause of this miraculous geometry is that the outer surface (the surface farthest from the axis around which the coiling takes place) grows faster than the inner surface, producing a geometry that is mirrored in a diverse array of forms—ram's horns, an unfurling fern frond, the eddies in a stream, even the meanders of Lewis Creek, which double back on themselves to fill space in the same manner as *Maclurites*.

There is another *spira mirabilis* on the island in the form of the dying leaves that cling to the oak trees on the south end of the flat one-acre island. While the maples and birches have lost all their foliage by December, these trees are more like the young beeches one comes across in the woods, holding fast to scores of warped brown leaves as if to say "no" to winter. Walking along the south shore of the island, where the limestone shelves out into the bay, these trees catch the eye. They look a bit like chestnut oaks, *Quercus prinus,* but they are smaller in stature, and the bark is lighter gray and flakier than the deep furrows of the chestnut oak. The leaves are narrower and the teeth on them sharper than what one's mind holds as its "chestnut oak" picture, too. These are yellow or chinquapin oaks, *Quercus muehlenbergii,* and the hundred or so mature trees on Gardiner's Island make up Vermont's only known yellow oak forest. Typical of dry woods on limestone soils, the tree ranges from southern Ontario and Wisconsin to southeastern Nebraska and northwest Florida, even creeping down into northern New Mexico, but its outposts in New England are few and far between. On Gardiner's Island it is joined by a dozen herbaceous plants also of southern affinity: the perfoliate bellwort *(Uvularia perfoliata);* rue anemone *(Anemonella thalictroides);* field meadow rue *(Thalictrum confine);* goldenseal *(Hydrastis canadensis);* a grass, rock dropseed *(Muhlenbergia sobolifera);* few-fruited sedge *(Carex oligocarpa);* small skullcap *(Scuttelaria parvula);* a bulrush, *Scirpus verecundus;* yellow pimpernel *(Taenidia integerrema);* two legumes—violet bush clover *(Lespedeza violacea)* and Canadian milk-vetch *(Astragalus canadensis).* A vine, hairy honeysuckle *(Lonicera hirsuta)* in some places twines over yet another "southern" plant—the shrub red mulberry *(Morus rubra).* Many of these plants are rare in Vermont, and some, like *Scutellaria, Taenidia,* and *Lespedeza,* are at their very northern range limit here.

Why does the yellow oak (and the group of "southern" herbs) flourish on Gardiner's Island? Once they probably flourished throughout the state, and even further north, and the island grove is a remnant of a more southern forest that thrived here some 4000 to 6000 years ago. This was the time of the Hypsithermal or climactic optimum, when annual temperatures were on the average a few degrees warmer than today. The yellow oak shared the landscape here with black gum, tulip poplar, sycamore, American chestnut, and other trees until around 1000 B.C., when the climate began to cool, and these species retreated along the major river valleys into southern New York and New England. Like many Hypsithermal relicts, *Quercus muehlenbergii* is now largely confined to the calcareous soils of the Champlain Valley. On Gardiner's Island the limy fossils have provided a refuge for these southern relicts; though calcium is essential to all green plants, too much is a bane, and there is often reduced competition on soils derived from limestone. With the other more typical oaks safely at bay, the yellow oak finds a footing on Gardiner's Island. It has kept its footing not only because of reduced competition, but because of the insular nature of the spot, which has protected the tree from its most indefatigable enemy—humans. While virtually every acre of the Champlain Valley's forests have been cut over repeatedly in the last few centuries, Gardiner's Island was too small to be farmed, and it was an inconvenient woodlot, so it has remained relatively free of the reach of ax and plow.

Before 1877, the tree was not known from Vermont at all, not because it didn't grow here, but because no one knew to look for it, or knew what it was when he saw it. The little grove on Gardiner's Island anonymously shed mast year after year like any other more common oak, or like the shagbarks that still grow there. Then on June 26, 1877, the Lewis Creek watershed's most persistent plant hunter paddled across Hawkins Bay to the cobble island. Cyrus Guernsey Pringle discovered what was to him known as the "Yellow Chestnut Oak" on that day, and after he came back in mid October to gather new fruits of the tree to be mounted along with the leaves and twigs he had collected in June, he never returned to the island. Other Vermont botanists—Ezra Brainerd, Willard Eggleston, Nellie Flynn, D. L. Dutton—knowing of Pringle's discovery, have trekked out to Gardiner's Island over the intervening years, but Pringle himself would each year be taken further afield, and the outpost oaks simply took their place in his botanical memory.

Pringle may very well have had oaks in mind when he first landed on Gardiner's Island. In 1876, Charles Sprague Sargent, director of the Arnold Arboretum, had set Pringle to search for *Quercus ambigua,* which was said by Francois Michaux to grow on the shores of Lake Champlain. The "ambiguity" of that tree was that it seemed to botanists to be a red oak *(Quercus rubra),* though it differed in some important characters. Sargent hoped to settle the question as to whether *Quercus ambigua* was a "good" species, or merely a hybrid or geographic variety of the familiar red oak. In searching for *Quercus ambigua,* Pringle began to acquaint himself with the tremendous variation in the native oaks of the Champlain Valley. He found groves of white *(Quercus alba)* and mossycup *(Quercus macrocarpa)* oaks, which also held many intermediate forms. He found red oaks that tended toward scarlet oak *(Quercus coccinea),* "pure" examples of which he discovered on the sandy bluffs and plains about the mouths of the Winooski and Lamoille rivers. He even found a "singular" tree standing between a white and chestnut oak whose bark was like that of a silver maple. He failed, however, to find Michaux's *Quercus ambigua.* In 1877 he continued his search, principally along the lakeshore, but of the "hundreds, perhaps thousands, of trees which [he] examined, *all were Quercus rubra.*" Pringle sent a selection of specimens to George Engelmann, the foremost expert on oaks in North America (it was Engelmann who described *Quercus muehlenbergii* in 1877 in the *Transactions of the Academy of Sciences of St. Louis*), noting the other oaks of the Champlain Valley: Mossycup and swamp white oak *(Quercus bicolor)* were "abundant"; chestnut oak (*Quercus prinus*) was "common on the warm rocky hills of W.Vt."; dwarf chestnut oak *(Quercus prinoides),* Pringle said (incorrectly, mistaking stunted young chestnut oaks for them), "extends to this vicinity," and he "knew of a few trees of *Quercus muehlenbergii* on an island of Lake Champlain." On Gardiner's Island he also stated that he had found trees that he could not decide whether to call *"bicolor or macrocarpa."* Engelmann and Pringle kept up their oak dialogue until Engelmann's death in 1882, bearing out a prophecy casually made in a letter to Pringle three years earlier: "Well, we shall never be done with the oaks, it seems!" Pringle himself once remarked that after twenty-five years of study he did not know the oaks of Vermont.

A dozen years after Pringle's trip out to the island, it appeared in a story by the Vermont writer Rowland Evans Robinson. "The Camp on the Lake" (*Sam Lovel's Camps,* 1889) is the tale of a young farmer, Sam Lovel; a younger and greener neighbor, Joseph Hill; Solon Briggs, a male Mrs. Malaprop; and their "Canuck" compatriot Antoine Bassette's June fishing excursion to the neighborhood of Hawkins Bay. These four men, along with Sam's wife Huldah; Uncle Lisha Peggs, a cobbler, and his wife Aunt Jerushy; Gran'ther Hill, Revolutionary War veteran and elder of the town; Mrs. Purin't'n, Huldah's mother; Pelatiah Gove, who later joins his comrades on this trip; and Joel Bartlett, a Quaker from "Lakefield," that is, Ferrisburg—comprise the central community of Robinson's fictional "Danvis tales." Robinson's fictional town of Danvis was set just beyond the Lewis Creek watershed and the adjacent Champlain Valley basins, up on the first ridge of the Front Range of the Green Mountains. Tradition says that Robinson based his fictional town on the hamlet of West Lincoln, which he had visited only once, with his father, when he was a boy of ten. It seems more likely that Danvis was an amalgamation of this and other adjacent Front Range towns—South Starksboro and Jerusalem, perhaps.

Whatever its source, the world of Danvis and its people appeared as Mount Abraham in Lincoln might look, viewed from the top of Shellhouse Mountain in Ferrisburg on a foggy day; its outlines were clear enough, but it looked distant, even though one knew it was well within reach. What put Mount Abe, West Lincoln—and Danvis—at a distance was not the fog, but the intrusion into Ferrisburg and the other valley towns of the modern era, with its telegraphs, railroads, and, close behind them, automobiles. Robinson, writing in the 1880s, set his Danvis stories between 1830 and the outbreak of the Civil War. In doing so he gently placed his fiction upon the land where modernity had not yet fully intruded—in an imaginary, unspoiled hill town of a generation before. Historians today would not rely on Robinson's stories for an accurate portrayal of antebellum Vermont, even though in the past some have enthusiastically done so. In 1936, John Spargo, President of the Vermont Historical Society, celebrated the historical value of Robinson's writings: "So realistic are these descriptions in fact that, if some great convulsion of nature, or some display of human madness triumphant, were to wipe all of Vermont out of existence, obliterating every trace of it, the discovery of a set of Robinson's books somewhere—perhaps on the shelves of some library in China—would make it possible for scientists to construct from them a faithful and dependable picture of Vermont as it was in Robinson's day. It would be possible to depict realistically the characteristic scenery, the homes, the occupations, the tools, the dress and the speech of the people."

Robinson himself gave a humbler explanation of his work. His prefatory note to *Danvis Folks* (1894) reads: "It was written with less purpose of telling any story than of recording the manners, customs, and speech in vogue fifty and sixty years ago in certain parts of New England. Manners have changed, many customs have become obsolete, and though the dialect is yet spoken by some in almost its original quaintness, abounding in odd similes and figures of speech, it is passing away; so that one may look forward to the time when a Yankee may not be known by his speech, unless perhaps he shall speak a little better English than some of his neighbors." The poet Hayden Carruth probably came closest to assessing the authenticity of Robinson's writings. In chronicling Danvis, Carruth said, Robinson aimed not for truth, but for "truth enlarged, the kind of superreal generalization at which all imaginative writing aims."

As necessary as it may have been for Robinson to settle his Danvis characters in a mountain town (as much out of the need to protect the identity of his Ferrisburg neighbors upon whom the characters were based, as it was for authenticity of speech and customs), his experience demanded that he bring them out of the hills to the marshlands of Lewis Creek, Little Otter Creek, and South Slang for his most evocative prose. "It was one of those lazy afternoons in June when all nature basks in the new warmth and nothing seems better to all things than to be still and enjoy laziness" when Sam

Lovel and his compatriots, atop a wagonload of camping outfit, arrived at the landing below the first falls of Little Otter. The year, though not stated, was likely around 1840. (Two passages in the book act as markers: "Some poor crazy creetur 'at orter be in Brattleburrer!" signals the earliest date, for the Vermont Asylum for the Insane opened in 1836; "A railroad in Vermont was undreamed of then" suggests 1848 as the endpoint, since in that year trees between Little Otter and Lewis Creek were felled for the Rutland and Burlington Railroad.) In 1840, camping was no fashionable pastime; in Robinson's words, "only white vagabonds and bands of Canadian Indians who had not much better shelter at home were supposed to live in shanties and tents for the pleasure of it, even in the pleasantest weather." The civilized valley folk who stared at the Danvis vagabonds were but two generations removed from a mandatory pioneer existence, and were loathe to seek it out voluntarily.

In response to their inquiries about good fishing spots, the miller at the first falls on Little Otter told them:

> "You c'n fish anywheres 't the's water 'n' ketch suthin' 'nuther, but 'f you want a ri' daown good campin' place, arter you git beyund the Slab Hole, you turn int' the left, on the wes' side o' the crik, op'site the san'bar, where the's a lot o' willers, an' you'll find the neatest place 't you ever see! Ye needn't build ye no shanty, for the's rocks a hangin' over 'at'll shelter ye, an' the's lots o' cedar browse tu make yer beds on, an' wood! The Slab Hole's full on't—lawgs, an' slabs, an' sticks o' fo' foot wood, 'n' everything, f'm kin'lin tu back lawgs. An' there ye be, right t' the lake, 'n' right t' the crick, an' Lewis Crik an' the seinin' graound not mor'n a quart' of a mild off!"

Robinson's faithful rendering of the miller's speech was the aural equivalent of his portraits of the local landscape; writing in dialect allowed him to let the place speak for itself.

Sam, Antoine, Joseph Hill, and Solon Briggs found the spot, decided to pitch a tent rather than use the rock shelter, but then were driven up to the rocky bluff above by hordes of mosquitoes. The carpet of fallen cedar leaves and moss that they found there led Antoine to exclaim: "Ah'll tol' you, boy, 'f he ant mek you felt sleepy for jes' look at dat beds!" The campsite was ideal, and served the Danvis party again in *Uncle Lisha's Outing* (1897). In Robinson's nonfictional writings, the Slab Hole camp appeared in 1894 in *Along Three Rivers*, and in 1895 in "A Voyage in the Dark," a story that appeared in the *Atlantic Monthly*.

Some idea of how frequently Robinson really visited the "Camp on the

Lake" can be found by looking at his field notebooks. Though a sketch dated October 1863 looks like a representation of the Slab Hole camp, it is unlabeled, and there is no accompanying textual material. The first notes about the camp seem to be an entry dated June 18, 1879: "Swarms of tiny fish about sand bar shore, transparent bodies, big eyes. . . . Found at 'Slab Hole' an arrow and a cutting tool, hatchet or chisel." On October 10 of the same year Robinson wrote, "loon by Gardiner's Island, his devilish laugh prophesying a wind. Joe thought he saw golden eagles sailing about Bluff Point. When we rounded the next headland green herons started from the tree tops. Found nearly a quart of pottery . . . at Slab Hole. . . . Flocks of ducks flying low looked as if on the water, skimming at rail road speed—a possible solution of the sea serpent stories." (Tales of a "lake monster" or "sea serpent" have been told by residents of the Champlain Valley for centuries; even Samuel de Champlain, who "discovered" the lake in 1609, told of one in his narrative of exploration.)

On August 20, 1882, Robinson explored the area around Hawkins Bay with Charles E. Faxon, America's foremost botanical illustrator: "Went to lake with Mr. Faxon. . . . Found what seemed like a low artificial mound west of Slab Hole." An avid amateur archaeologist, Robinson must have been anxious to excavate the mound, holding as it did the possibility that the Slab Hole campsite was once the habitation of more ancient hunters. It was not until almost a year later that he had his chance to investigate. In an entry dated August 16, 1883, Robinson writes: "Down the creek with my little girl. . . . Dug into Mr. Hamner's mound—found nothing but old leaves bedded in hard clay—a 'cradle knoll,' I think. Pickerelweed blooming in the marshes and cardinal flower blazing in the lowlands."

Pickerelweed still blooms in the "wide ma'sh" at the mouth of Little Otter, and purple loosestrife lines the cream-colored sandbar that marks the Slab Hole, where each spring new flotsam piles up at the base of the limestone bluff. The mosquitoes are still fierce there, and you can stand in the shallows near the sandbar and feel hundreds of transparent fry, moving out from their nurseries in the cattail marsh, pass alongside your submerged legs. The spot has just the right balance of prospect and privacy to invite transient settlement. The Slab Hole camp was Rowland Robinson's camp, shared perennially with his friends Joe Birkett and Sedgwick Preston, and his cousin Rowland Robinson Mintern. If you poke around in the cedar duff, you might find the refuse of their times together here—charcoal from their campfires, empty tobacco tins, perhaps even a clay pipe

Slab Hole camp of Rowland Evans Robinson's fictional "Danvis Tales." Courtesy Rokeby Museum, Ferrisburg, Vermont.

or a fishhook. That fraternity was transformed by Robinson's imagination into the Danvis hunting and fishing expeditions. Instead of the long trek on winding dirt roads from beyond Hogback Mountain, they had only to come up the East Slang, on whose northern shore their skiffs were always moored, to Little Otter, a journey of less than a mile to reach the Slab Hole and Hawkins Bay.

Rowland Robinson never called the bay into which Lewis Creek empties "Hawkins Bay"; he referred to it by the name the French explorers had given it—the *Baye des Vaisseaux*. For more than a century and a half after the first French exploration of Lake Champlain, it was still the "Bay of Vessels." British naval officer Captain William Chambers, who surveyed the lake to help the Royal Navy navigate it during the Revolutionary War, called it such, and his description seems to favor the bay's suitability for anchoring large ships as the true meaning of the name:

> As soon as you pass Split Rock in going up the Lake, on the East Shore is Baye des Vaisseaux which I think is one of the best Bays on Lake Champlain you may Anchor in any Water from thirty fathoms to two but what I think is the best birth is about three Cables length to the WNW of a small Island [Gardiner's Island] that lies off Little Otter Creek where you will have nine Fathoms if the wind should blow very hard from the Northward, you may Anchor in the NE part of the Bay in four, five, three or two Fathoms Water muddy ground about half a Mile from the Shore, in short you may trust to your Lead for the Soundings as very regular, and most of the Bay has a muddy Bottom.

Robinson's inquiries could never satisfy him as to the reason for the French appellation: "whether because of the earthen pots or *'koks'* of the Indians which they may have found there, and whose rudely figured shards are yet strewn on the sandy shores, or in honor of their own water-craft, one can only guess and decide by one's own fancy. It has no name now save so far as its many indentations bear the shifting names of adjacent land owners."

The naming of places for people miffed Robinson. Grounded as he was in history, he recognized such commemorative names as transient and meaningless. "This changing of names with ownership is often confusing and always belittling, as in the case of Thompson's point, which name, though long established, is weak and insignificant compared with its fine old Indian name, *Ko zo-wa ap-ska,* the Long Stony Point. Such a point it always was and always will be, whereas it was some Thompson's freehold but for a few years at most, and Thompson himself was long, long ago, dead and forgotten."

Rowland Robinson's longing for things aboriginal went beyond place names to the artifacts of Abenaki existence. He was a "pot hunter," but not in the typical Victorian mode of displaying lifeless relics far removed from their original context. He sought arrowheads and pottery shards to make real the ghosts of the people who once lived there. He longed for the Bay of Vessels to give him "one perfect succotash-pot just as it came from the hand of the Waubanakee squaw that fashioned it, or with the smutch of camp-fire smoke upon it. I should prize it above all the old china in the world. But I was born too late for such a gift, and get only shards."

The sand bar at the Slab Hole where Robinson got "only shards" belongs properly to Little Otter, though its construction owes not a little to the sediments of Lewis Creek as well. Lewis Creek has its own sand bar, actually the emergent portion of a fan-shaped delta at the Creek's mouth.

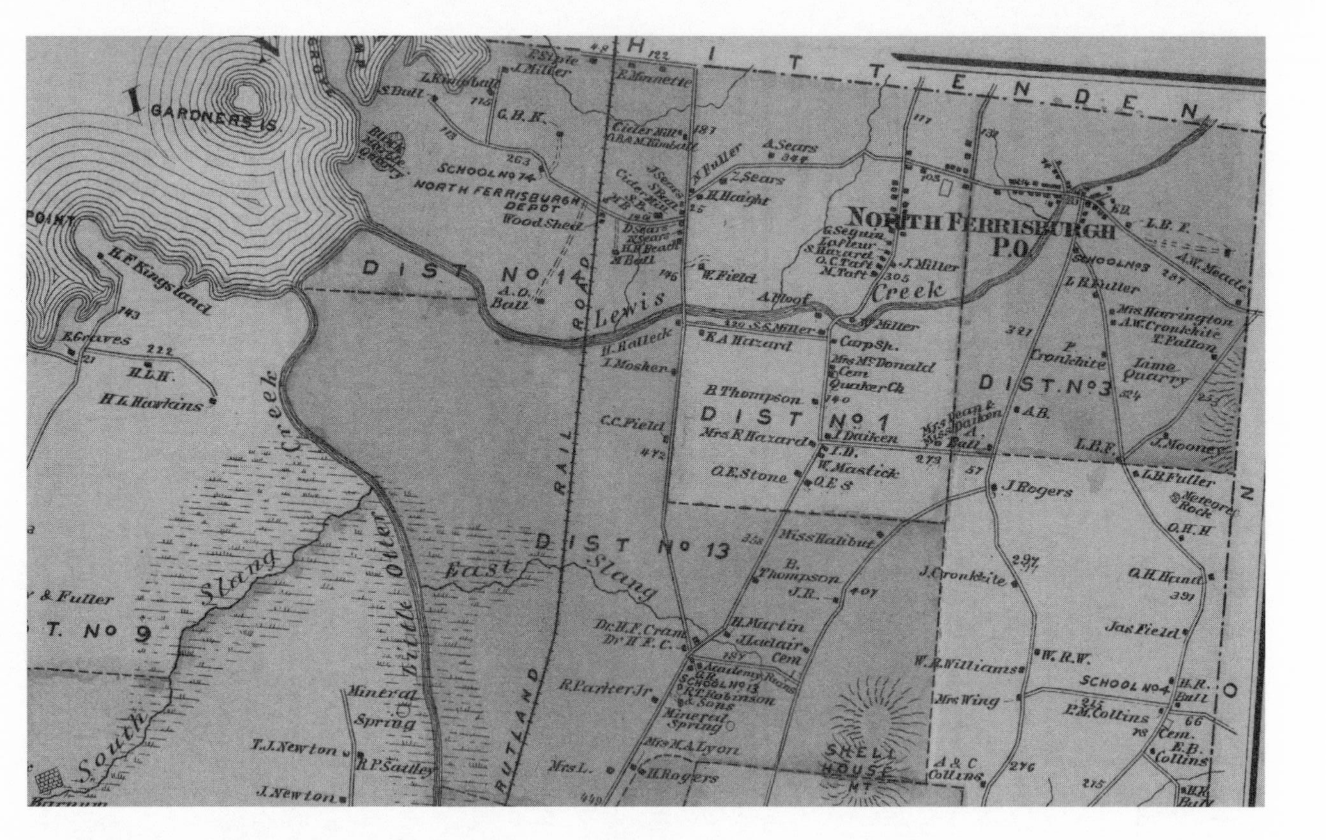

From F. W. Beers, *Atlas of Addison County, Vermont* (New York: F. W. Beers, A. D. Ellis, and G. G. Soule, 1871).

The sand connects a small rocky island to the marshy mainland, making the landform in strict geomorphological terms a "tombolo," but for the first two months of each spring the point remains an island, its sandy connection submerged by both the lake's and Lewis Creek's high water. As dynamic a landform as most river deltas are, this one, in the last century at least, has changed very little. Rowland Robinson described it in 1889:

> Before them a long incurved beach stretched away to the north, ending at a rocky point. The waves of immemorial years had thrown up the sand into a low breastwork wherein flourished a rank growth of rushes, sedges, and other aquatic plants, nourished by the undisturbed muck of their own decay. So close along the waterline that their wave-washed roots were spread like a tangled net upon the sand stood an irregular row of great water maples with tower-like trunks, buttressed, loop-holed, mossed and lichened by age, scarred by the battering rams of ice that the lake had hurled against them, with tops wind torn and decaying, but sending up new smooth trunks and abroad with youthful vigor a graceful ramage of branches and fresh leafage as if they might endure a thousand years.

Robinson's description belonged to his fictional *Sam Lovel's Camps* narrative, and represented the sand bar circa 1840. He could not refrain from concluding his description with the interjection of the 1889 picture: "They are gone now, and their ancient sites are marked only by rotting stumps on the barren unshaded shore. A meaner and deadlier foe than time, or wind, or waves has sapped their foundations, and years ago they were peddled out at so much a cord by their avaricious owner, who begrudged even the shadow of a tree." A century has healed this wound; today the sand bar is again clothed in silver maples, younger than the "great" ones that greeted Sam Lovel, but as ice-battered and wind-torn, with fully equal prospect of reaching "mossed and lichened" old age. The rocky point now sports a summer camp, but one feels it may be less permanent than the silver maples and beach wrack along the sandy shore.

To local fishermen the sand bar was the "seining ground," and again one must turn to Rowland Robinson to capture the sights and sounds of a century ago. The Danvis men, expert anglers though they were, "could not fish for all Danvis nor the half of it," so they set out from the Slab Hole camp one day to barter with the fishermen who were hauling their nets near the mouth of Lewis Creek.

> There were two gangs of seiners on the beach. The three men composing one gang were Canadians. . . . Their chief was an old fellow of large build,

of greatest dimensions at the hips, tapering thence upward to his ears and downward to his bare feet. It was from the interior of this widest region, apparently, that his broken English was laboriously upheaved to the surface with intermittent guttural grunts. His face bore a grim expression of good nature and also a pock-marked red nose that resembled in shape and color an immense strawberry. His younger assistants, who were clearing the net of sticks, weeds, and clams, and folding it onto the broad stern of their scow, appeared to be his nephews, for they frequently addressed him as Onc' Theophile.

"Haow de du?" Sam saluted him.

"Ough! How do," Uncle Theophile grunted in labored response, and then glibly gave in French an order to one of his nephews.

"Hevin' any lick tu-day?" Sam inquired with an assumed languor of interest.

"Make, ough, one haul, ough; gat dat," Uncle Theophile answered, pointing to a bushel basket half full of pike perch and pickerel.

"Wal, that'll du tol'lable well," Sam said after tilting the basket till some fish were exposed and critically examining the gaping mass, "haow much be you goin' tu tax us for, wal, say four haul?"

"Ough, twenty-fav cen' haul." Theophile answered, coiling the elm-bark seine ropes on the beach, "fo' haul, ough, dollar."

Antoine succeeded in bringing the exorbitant price of a single dollar down to 80 cents by offering to help his countrymen with their net, and Robinson goes on to describe in detail the methods of the seiners. The boat headed out from the sand bar beach in a long curve toward the rocky islet, the net slowly draped behind until the last "tommy stick"—the wooden staves that spread the ends of the seine—went overboard:

Then the scow headed for the beach, trailing out the bark rope till she grounded, and the crew, tumbling out, began to haul on it. Antoine, now an obedient assistant, hauled with Theophile on the other rope, while the old man gave out concise orders.

"Tirer! Tirer!" or *"Doucement! Douce-ment! Tirer pas ca vite!"* as occasion required.

Presently the tops of the tommy-sticks appeared at the ends of the approaching curve of floats that rippled the water with a hundred wakes, and then as they climbed the long slant of the bottom and showed half their length inclined inward, one of the nephews dashed out and gave the stick at their end its proper outward pitch, while Antoine in unquestioning response to Uncle Theophile's command, waded out mid-leg deep to perform the same office for theirs.

The water inside the net was now boiling with struggling fish, and the ropes were tossed with frequent splashes to frighten them back within the narrowing barrier, over which now and then some desperate captive would leap and regain freedom. Sam thought that in these instantaneous flashes of gleaming scales and glistening water droplets he recognized the forms of bass, and could not help feeling glad that such gallant fish had escaped such ignominious capture. But even his love of fair play could not withstand the excitement of so good a haul, and now that the ends of the net were landed, and it was hauled steadily in till the bellying bag stranded its writhing and grasping burden, he was as busy as the others tossing out pike-perch, pickerel, bass, suckers, mullet, perch and sunfish that glittered on the gray sand in a great heap of mother-of-pearl, emerald, silver, and gold.

"Dah, seh!" said Antoine, proudly, when the net was emptied, "ant Ah'll mek it pooty good hauls?" Ah'll de boy can ketch de feesh ev'ree way Ah'll man to ketched it! De hookanline, de spear, de nets, Ah'll gat no different of it me!"

Robinson's opinion about the practice of seining is partially voiced by Sam Lovel, who leaves the seining scene to paddle his canoe up Lewis Creek in search of "'Swago"—black bass:

"It's a good 'nough way tu get fish, but 't ain't no gre't fun fur me. The best part o' fishin' is lackin'. The' hain't no fair play 'baout it, an' it makes me feel kinder mean."

"Wal, naow, Samwell," said Joseph, pondering, while he searched for his pipe in every pocket but the one it was in, "seems 's 'ough 'f I was a fish, an' it mos' seems 's 'ough I was, a-drinkin' nothin' but water, 'at I'd livser be swe' up kinder easy in a net wi' a hull lot for comp'ny in misery an' tu be fooled wi' a worm or suthin' wi' a hook inside on't, an' then hev my jaw half tore off, julluk ol' Darkter Wood pullin' a back tooth."

"I wa'n't considerin' on't f'm the fish side," said Sam, "but fish does hev jes' much fun a-foolin' us as we du them. Why I've seen an ol' Beav' Medder traout laugh clean tu the end of his tail when he'd peeled my hook bare naked, an' I b'lieve them 'ere 'Swagos is up tu jes' 's much fun 's a traout is."

Robinson's distaste for seining went beyond the feeling that it lacked sport to an abhorrence of a method that so seriously depleted the available stock of fish. In 1840, the approximate date of the story, seining was a completely acceptable form of angling, but at the time when Robinson wrote *Sam Lovel's Camps,* it was largely outlawed, in no small part due to Robinson's efforts. Fish and game laws were a new and untried thing then, and the Vermont legislature seemed as preoccupied with them as with railroads and prisons. In 1876, the first real attempt at enacting a comprehensive set of fish and game laws was made, but little attention was given to the destructive practice of seining. That same year Robinson persistently lobbied Fish Commissioner M. C. Edmunds to halt seining. Robinson believed that all netting in streams and within a certain distance of their mouths should be entirely prohibited: "The wholesale destruction of our Bull Pouts should be stopped—they have been so unmercifully seined here, that the majority of them caught this season are poor little fingerlings." One indication of the lack of a widespread conservation consciousness can be seen in another law passed in 1876—one to prohibit the use of explosives for catching fish (fine—$25). Two years later a five-year moratorium on seining was put into effect, setting a fine of $10 for violators; however, the law provided for six weeks of net-fishing between October 1 and November 15 on Lake Champlain and its tributaries. There were no enforcement provisions in these laws, so they were broken regularly with impunity, and Robinson corresponded frequently with the state Fish Commissioners both to inform them of infractions and to urge them to ask the legislature for tougher laws and more enforcement power. In 1882, the law curtailing seining was changed to include "all pound net, trap net, gill net, set net, and fyke fishing" and the fine was raised to $100. Four years later a bounty of $10 (for nets of less than 5½-inch mesh) was added as inducement for citizen enforcement. The illusion of a radically aware public and legislature is swept away if one looks at other 1886 fish and game legislation: That year a bill was also passed that upped the bounty for wolf, panther, and bear to $6, lynx to $5, and fox to 50¢! In 1892 the Vermont fish and game laws were extensively revised, including the seining restrictions, and in 1894 the legislation had been further refined (or watered down, depending on your point of view) to allow seining in Lake Champlain and its tributaries between October 1 and December 1, by permit. In 1896 the antiseine law was broadened to include a $100 fine for the *possession* of nets, and Robinson was influential in seeing other progressive legislation passed that year—an act prohibiting the use of nets, snares, and jack lights in hunting duck.

The Hawkins Bay haunts reappeared in Robinson's 1897 collection, *Uncle Lisha's Shop*. This time Sam Lovel is joined by Joseph Hill, Uncle Lisha, and Antoine for a fall duck-hunting expedition. One day toward the end of their encampment the men are joined by Sam and "Lisher's" wives, Huldah Lovel and Jerusha Peggs, and the two couples go fishing along the shore of "Garden" (i.e., Gardiner's) Island. After Jerusha has caught a pickerel off the gray northern wall of the island, they land, and, with Sam acting as guide, explore its "interior." Robinson verifies the island's fictitious name by citing the "garden-like bloom of its shrubbery . . . the abundant black clusters of viburnum berries and scarlet haws of wild roses, and . . . blue and white blossoms of asters."

The fishing party does its own bit of hunting on the island, not for rare oaks but for Indian artifacts that will fetch them back to a past only recently disappeared. During their search, they come across a different sort of excavation—a "money diggers'" pit. The money diggers were anonymous townfolk, who, fed by tales of fabulous contraband treasure buried by smugglers on Lake Champlain, excavated in hopes of finding it. Although perennially the hunter, Sam's quarry was never sought for material gain, so Sam himself was no money digger. He preferred to dig for what Thoreau called "fossil thoughts." On the narrow strip of gravelly beach on the east end of the island, they find arrow points of flint, left behind by some earlier, aboriginal, duck-hunting party. Admiring the craftsmanship of the points, Uncle Lisha remarks, "It does beat all natur' haow the critters made 'em! We could n't, wi' all the tools we got, an' I hearn an ol' feller tell aout West 'at the Injuns done it wi' a sort o' bone thingumajig, jest by pushin' on 't with the hand, an' he claimed he'd seen 'em at it, but I d'know 'baout it. That 'ere 'd mak a toll'able good gun-flint, an' I guess I'll keep it." And with that the bit of silica plucked out of a limestone bed by a long-dead Abenaki hunter, who shaped it into its lethal form to take a teal or canvasback unknown centuries ago, became the stuff to spark an old Yankee's hunting piece. Robinson drew upon his vast store of local knowledge to spin the treasure-seekers' pit, the piece of flint, the limestone shelves, the canvasback and teal, and his imaginary Danvis characters into an intricate web of relation to the past.

The web of fact and fiction in Rowland Robinson's stories can perhaps best be unraveled by examining his diaries. Robinson's "diaries" are really field notebooks, filled with brief observations about wildlife, the weather, folklore, and local history. They were never systematized or complete, but rather a "collection of days." Their interest lies in the fact that they are as much a map of Robinson's psyche as his books are. They are the bare bones of his finished work, in the same way that the pencil sketches in them are the material from which his watercolor and oil paintings or ink drawings were derived. In a general way, every entry seeks to pinpoint some happening, some small event; behind them all is a vague feeling that these events are being marked for some future purpose, stored away so that they might be recovered for use at a later time. Historical hindsight alone might generate this feeling, for anyone who is familiar with Robinson's published work will see page after page of root material—places, people, events—for his stories. Occasionally there are entries whose tone emphatically states that they *are* observations to be recalled and reused. On August 12, 1880, Robinson wrote: "Fishing at Garden Island, with Joe Birkett and R. N. Preston. *South of first [illegible], in line N. with w. point of [illegible], end of Long Point and a barn. East with oak on south end of little island at mouth of Lewis Creek, and a little north of first great mtn., south of but nearer than Camel's Hump*. Caught 15 bass, 4 pickerel and several perch." Using both proximal and distant landmarks, Robinson the fisherman was fixing his position for a future fishing trip. Abroad on Hawkins Bay, unable to poke a tally stick into the mud as he might do if he were hunting muskrat in the marsh, he marks the spot by an inexact but accurate enough triangulation.

In a sense, Robinson's entire artistry was just such a series of triangulations, between the points of place, people, and events. Place, fixed and firm despite man's alterations, was most important; it was the benchmark for Robinson's suryeying. People and events shifted during his lifetime, but he did his best to fix them. That effort began in earnest in 1858, when Abby Hemenway enlisted Robinson to contribute a sketch of Ferrisburg's pioneer history to the first projected volume of her five-volume *Vermont Historical Magazine*. Since the early records of the town had been destroyed by fire in 1785, Robinson had to rely on other sources: He interviewed descendants of the first settlers, examined old cellar holes, located mill sites, and traced abandoned roads. In his notebooks, there are no alidade readings, but page after page of a different type of raw data: wildflower blossoming dates, the condition of creeks, the status of fish and wildfowl populations, the siting of muskrat houses. There are historical data too—place names, tall tales, outdoor lore—from the elders of Ferrisburg. Less commonly, there are readings of his own temperament, of moments of elation or despair or of flights of his imagination. One such fancy came to him on the same day as the bonanza fishing trip triangulated in his notebook: "Split Rock Mtn. dim with haze, the water line cutting straight and dark against its base. A war party of Abenaki or Iroquois might slip past close to it, unseen by us, *behind the round of the world*." In 1884, Robinson's story "On a Glass Roof: Ice-Fishing on Lake Champlain" was published in the *Burlington Free Press*. He tells how he goes out one March day to Hawkins Bay to try his hand at ice-fishing, and there meets "a Waubanakee of St. Francis, plying the gentle art here in the war-path of his ancestors . . . from our low stand-point the rough, indented shore of Split Rock Mtn. showed only as a straight line, and it seemed as if a war-party might slip by unseen behind the round of the world." This is but one example of Robinson drawing directly on his notebooks not just for the purpose of jogging his memory, but for the exact prose of that memory's moment. It is a particularly apt passage though, for it embodies the uniqueness of Robinson's vision. Everywhere he looked, just over the horizon there lingered the shadows of things gone before.

Rowland Robinson knew deeply the timelessness of place, knew it as well as he knew each bend in his beloved "Sungahneetook"—Lewis Creek. All of the imaginary happenings of the "Danvis tales," the semiautobiographical stories that make up the bulk of Robinson's fiction, seem to be held lightly but firmly against that delicate gossamer web of time. The chapter that brought the Danvis folk onto Garden Island closes with Robinson's paying homage to Time. Sam and Company abandon the bluff camp to head back to their mountain hamlet of Danvis, perhaps to return some other October to the spot, perhaps not. Taking their trophies—pickerel, teal, and a handful of arrowheads—with them, they leave only the campfire behind: "The last ember snapped out in dull explosion and the last thin wisp of smoke dissolved in the colorless air, and amid the silence of desertion the falling leaves began the slow obliteration of man's transitory sojourn."

Time's obliteration of humanity's transitory sojourn thankfully is a gradual process, leaving traces for each generation to discover of their predecessors' campfires. The three historical figures chiefly considered here—Perry, Pringle, and Robinson—were each, like Sam Lovel, hunting for just the right place to make their camp. They demanded different features of their surround: Cyrus Pringle sought the full complement of his natal watershed's floral diversity, as part of a wider quest for botanical knowledge; Rowland Robinson poked about Lewis Creek's past in search of the sites and stories that best made his fellow moderns rediscover their common humanity through the experience of place; John Bulkley Perry dug deep into the region's geologic past to anchor the land in the nineteenth century's unfolding narrative of earth history. Though during their lifetimes they never walked together any stretch of Lewis Creek, they walk together now, brought into contact by time's passage, as their biographies begin to speak to us here at the end of the twentieth century.

Those biographies speak to each other and to us because they were engaged in an endeavor that has only grown more challenging in the century that has passed since Rowland Robinson's death in 1900. Perry, Pringle, and Robinson plunged fully into the physical world, exhilarated by their individual powers of exploration. They lived in an age when all America shared that conquest of the physical world. They also shared with all America the quest for redemption in the more-than-physical world, at a time when that world began to lose its potency. With each step toward a more precise delineation of the physical landscape, an older, invisible landscape of spirit receded from view. Though they could each easily set up comfortable camps upon the banks of Lewis Creek, striking their tents upon the fog-shrouded shore of Heaven had become a daunting task. It is nonetheless daunting today; indeed, the fog grows heavier, while our desire to strike land is unabated.

CHAPTER 2

Liminal Places and People

I enter a swamp as a sacred place. . . . A town is saved, not more by the righteous men in it than by the woods and swamps that surround it.
—Henry David Thoreau, 1861

Swamps— These are hardly of sufficient importance to deserve a separate notice . . . When the country was new, there were many stagnant coves along the margin and among the islands of Lake Champlain, which, during the hotter parts of the summer, generated intermittent and bilious fevers. But, since the clearing of the country, these have been, to a considerable extent, filled up. . . .
—Zadock Thompson, 1849

IF ANY SINGLE habitat could be said to have suffered most from human nature-hating, it would be the swamp. Zadock Thompson was a keen naturalist, yet his attitude toward swamps, as witnessed by his remarks just given, was one of disdain. The average citizen of mid-nineteenth-century America would have made an even more vituperative evaluation; that citizen held the local swamp in about as high esteem as rattlesnakes, wolves, and chicken hawks. The prejudice against swamps drew its strength from two sources: the inconvenience they presented to a nation of farmers bent on tilling every available inch of soil; and a deeper, almost archetypal fear of such places. Swamps were often impenetrable and always inscrutable. They were the part of Eden where humans were not welcome, though a host of other creatures were welcome there. There was a great stirring of life in a swamp, a rank fecundity that spoke of birth equal with death and decay. If the theory of spontaneous generation still had its adherents in the mid nineteenth century, they likely pointed to the swamp. Things came out of the swamp. Some contemporary Vermonters believe that "Bigfoot"—a ten-foot-tall, musky-smelling monster whose blood-curdling shriek can be heard for miles—lives there. To back up such accounts, cryptozoologists cite the nineteenth-century reports of a similar creature called "Old Slipper-Skins," and note that the Abenaki whom Samuel de Champlain encountered around the magnificent Missisquoi Bay swamps told of *wejuk,* a creature that loosely matched Bigfoot's description.

The wooded swamp at the mouth of Lewis Creek owes its existence to both the Creek and Lake Champlain. Downstream from about the railroad bridge, the Creek has carved a substantial floodplain in the clayey soil, especially on the south side of the Creek. At its southeastern edge the floodplain shows its riverine origin in the sinuous scalloping of the land where the channel used to run. A pair of natural levees bordering Lewis Creek effectively separates the stream from the adjacent swampland. Following the annual spring ice-out of Lake Champlain, however, the swamp forest is flooded by the swollen lake and continues to be inundated for much or all of the spring. In early April one can canoe out the mouth of the Creek, turn south and east over the tops of last year's cattails and wild rice, then make one's way back into the flooded forest. The swamp is like a Georgia bald cypress swamp then, though instead of bald cypress the trees are silver maples, the hybrid maple *Acer* × *Freemanii,* green ash, swamp white oak, and elm.

Usually by late May or early June the waters have receded from the swamp, leaving the damp silts bare to receive a rain of newly matured seed, which germinates rapidly and perpetuates the swamp forest. In some years, however, the lake level remains above 97 feet through early summer, and no seedlings take root. In those years the mature trees, except for the ones already stressed or dying, are not harmed. Similarly, periodic short-lived flooding in other seasons has little effect on tree health. However, two consecutive seasons of inundation causes a loss in vigor and some die-off in silver maple. Green ash, more flood tolerant, succumbs after three years, and by the fourth consecutive year of flooding lasting for more than half of the growing season, all woody vegetation dies. Such a dieback occurred most recently in the early 1970s, when all trees below 96.5 feet were killed due to high lake levels that continued from 1969 to 1974.

Just below the swamp forest to lakeward is a zone of buttonbush and red-osier dogwood, and then the woody plants of the swamp give way to the herbaceous marsh. Here, due to the nutrient-rich sediments that have been carried by Lewis Creek from its far-flung tributaries, even the fertility of the swamp is surpassed. There is a green explosion of littoral plants rooted in the detritus brought from the entire watershed. Cattails (*Typha* spp.) and wild rice (*Zizania aquatica,* in the past commonly known as "wild oats") are the tallest and most conspicuous members of the marsh community, but there are a host of other species that contribute to the enormous biomass of this rich ecosystem: broad-fruited bur reed and broad-leaved arrowhead, whose tubers are favorite foods of muskrats and ducks; river bulrush and a suite of other bulrushes—wool-grass, slender bulrush, and hard-stem bulrush. These are the "emergent" plants, whose thick, flexible stems and bladelike leaves form a forest in miniature at the interface of land and water. Within that forest thrive an incredible menagerie of animals; clams, worms, snails, insects, and fish populate the underwater portion, while above water, ducks, rails, herons, terns, blackbirds, wrens, muskrats, and mink feed and breed among the plenty and safety of the marsh vegetation. Though the Lewis Creek marsh proper is only about 315 acres in extent, together with the wild rice marsh at the mouth of Little Otter it makes up over a thousand acres, and is considered the finest large expanse of marsh in Vermont.

The marsh appears one-dimensional to our eyes, yet it is intricately and densely three-dimensional. The cattail stands are punctuated by potholes and laced with a maze of narrow waterways filled with floating or submerged vegetation: coontail, greater and lesser duckweed, milfoil, waterweed, and a dozen species of pondweed. Yellow pond lily, white water lily, and pickerelweed add splashes of color to the green riot. Aquatic insect larvae everywhere move from their watery nursery up emergent stems to join the extra-aqueous world. Red-winged blackbirds, swamp sparrows, and long-billed marsh wrens sing from the summits of the marsh forest—the tops of cattails—while rails and bitterns stalk silently below. Above it all hang the harrier and the short-eared owl, the only two avian predators who can exploit the resources of the watery expanse.

Of all the abundant wild life of the marsh, Rowland Robinson wrote most frequently about the ducks. His interest in waterfowl was representative of his neighbors of his own era, as well as of generations before and after. Every fall, duck blinds in the Lewis Creek marsh that have done service for decades are refurbished by hopeful gunners. In Robinson's father's day, the duck hunting was done strictly by local hunters and an occasional Abenaki down from Canada; in his own day there were added a group or two of hill folk (models for Robinson's "Danvis" men); today the painted wooden slabs identifying the blinds' owners bear the names of hunters from all over the Champlain Valley and beyond. The fall migration has always brought a wide variety of ducks; in the marsh there are widgeons, pintails, shovelers, gadwalls, green- and blue-winged teal, black ducks and mallards, while on the open water of Hawkins Bay and the mouths of Lewis and Little Otter are goldeneyes, bufflehead, ring-necked ducks, and greater and lesser scaups. The duck hunters' most frequent quarry was and still is the common mallard and its close relative the black duck, known in Robinson's day as the "dusky duck."

After these two the most abundant duck to be found in the Lewis Creek/Little Otter Creek marsh is the blue-winged teal, whose population was annually fattened on the wild rice crop there. Robinson recorded his own first encounter with this little duck by way of describing Sam Lovel's:

> Antoine bent his head down low as a flock of teal came stringing down the channel in arrowy flight, and Sam, aiming a little ahead of the leading bird, pulled trigger. The hindmost teal in the line slanted downward, and, striking the water with a resounding splash, lay motionless when the impetus of its fall was spent.
>
> "Wal, if that don't beat all natur'," Sam said with a gasp of surprise. "That 'ere duck was ten foot ahind o' the one I shot at. What sort o' ducks du ye call 'em, Antwine?"
>
> "He come 'fore you call it dis tam, but w'en he ant, you call heem steal dawk in Angleesh, Ah b'lieved so. He was plumpy leetle feller," Antoine remarked as he picked up the bird, when Sam had reloaded and the canoe was again in mid-channel.
>
> "An' a lively breed they be, tu shoot a-flyin'," Sam commented, as he examined this victim of chance. "'T ain't no use a-shootin' at 'em. You got to shoot 'way off int' the air ahead on 'em, an' let 'em run ag'in your shot."

Not far from the fictional spot on Little Otter where Sam got his first teal, young Rowland Robinson got his first duck, though it was most likely a mallard or "dusky." It was an important rite of passage for the boy, who, as he wrote of the event in his twilight years, did so in an extended plea for "hunting without a gun":

> Rounding the bend, half-way between the Myers Landing and the Sattley Landing, I come to the turn of the channel that I can never forget while I

Sketch of wood duck by Rowland Evans Robinson. Courtesy Rokeby Museum, Ferrisburg, Vermont.

> remember anything of the stream, for here I killed my first duck, shooting it on the wing, astonishing myself no less than Jule Dop, who paddled the boat for me. It was enough glory for one day to have that matchless paddler regard me with unfeigned admiration, and he not less than three years my elder. . . .
>
> If I might by any shot at anything, once more have my heart warmed with such exhilarating fire as that shot set aflame in it, I could not with any sincerity recommend this blood-guiltless hunting, nor practice what I now uphold.

The wood duck, however, captivated Robinson the hunter most completely. It may have been his artist's eye, appreciative of the male wood duck's spectacularly iridescent breeding plumage, which made this his favorite. The woods at the mouth of Lewis Creek, particularly those ringing "Jigwallick" (an Anglo version of the Abenaki *chegwalék,* meaning "at the place of the frog"), the lagoon that runs back parallel with and south of the Creek, were and still are prime habitat for wood duck, furnishing a continually replenished crop of dead trees for nest sites. (Today the snag and stump cavities are augmented by nest boxes placed along the Creek by the Fish and Wildlife Department.) More than any other creature, Robinson lamented its disappearance from the woods and waters of Lewis Creek. Around the turn of the century there was widespread concern that the wood duck might become extinct, due to both an excessively long season (eight months) and extensive habitat destruction. The Migratory Bird

Treaty Act of 1918 helped stave off extinction, and restricted seasons during the 1920s and 1930s led to a dramatic recovery of the wood duck. Listed for Vermont as a rare summer resident in 1907, in 1971 it was estimated that the species had a breeding population of around 15,000 in the state.

The hunter's vision of scarcity and fear of extermination, not some deep-seated reverence for nature, was the prime mover for early conservation efforts. Even Rowland Robinson, who usually championed the cause of wildlife for nonutilitarian reasons, sometimes seemed most melancholic about the decline of the wood duck because it afforded him fewer targets. In "Given Away," he tells the story of his discovery of an old oxbow slough in the midst of the swamp woods of Lewis Creek:

> Its surface was stirred by something which I could not see moving upon it, and I crept cautiously to a point that gave me a view of almost its whole length. What I beheld nearly took my breath away. The little lagoon swarmed with wood ducks, some in rows on the many mossy old logs that lay athwart and along it, some comfortably asleep, with head indrawn or tucked under a wing, some preening their gay plumage, some standing upright to stretch their wings, while the water was alive with others, indolently swimming to and fro, seaming the duckweed with innumerable aqueous paths, or nibbling the water, or thrusting their heads beneath it, and all in abandonment to a perfect sense of security that it was cruel to disturb.
>
> No emotion of pity softened the youthful savagery of my heart. It beat only with the joy of great discovery—the chance of a lifetime that lay before me. It beat so vehemently that it is a wonder I even hit the pool, to say nothing of hitting one of the uncounted dozen of ducks ranged on the nearest log, for whom my aim was intended—yet I saw three tumble helplessly from their perch, and when with a roar of wings that was like a prolongation of the report of my gun, innumerable ducks arose and filled the air before me, I fired wildly into it, two more chance-stricken victims of the aimless shot plunged back into the troubled water. The ducks seemed unable to realize that this safe retreat had been discovered and invaded by a cruel, relentless foe, for they continued to circle and hover over it till, with trembling hands, in more haste than speed, I reloaded my gun, and, grown cool enough to select single birds, brought one down with each barrel.
>
> Then the last and boldest lingerer reluctantly departed, and the silence of desertion fell upon the place, except as I splashed and poked about it to secure my game; and, with a view to future onslaughts, made a path for a stealthy approach, clearing away every sprout and dry twig that might swish or snap a signal of alarm. There was not a sign to show that the place was ever visited by any one else, and I congratulated myself on possessing sole knowledge of its existence.

Many a day thereafter I went to it alone, guided from afar by the oak and pepperidge, which towering above the second growth, were unmistakable landmarks, whether in leafage of green or scarlet and brown, or in gray nakedness. While I kept my secret, seldom was a visit unrewarded by at least one shot at wood ducks, or later in the season at the larger and warier dusky ducks, which haunted the sequestered slough until it was frozen.

But in an evil hour I disclosed it, under promise of secrecy, to a faithless friend after an unsuccessful day with him on the two creeks. It was not long before the path was worn by the frequent tread of other feet than mine, and ducks began to be shy of a retreat that no longer promised rest and safety. In two years it was common to every gunner in the neighborhood, and worth no one's while to visit.

As one still searches for something lost past all hope of finding, so was I now and then drawn thither, but never to find more than a solitary heron standing like a gray statue in the desolate slough, or a lone sandpiper skirting the low shore, or perchance a muskrat channeling the duckweed with his silent wake. I had given away my discovery only to have it made worthless.

Wetlands the world over have had their memorialists, writers who tried to portray in words the overwhelming life they found in places between water and land, and so it is with the Lewis Creek marsh. It was immortalized in Rowland Robinson's fictional and nonfictional work, brought to readers first on the pages of *Forest and Stream* and then in books. Along with the Little Otter marsh it was the scene of innumerable experiences of the Danvis men on their forays to the Champlain Valley. Perhaps the first description of it appeared in 1889 in *Sam Lovel's Camps*. Sam, Antoine, Joseph, and Solon have just set out from the Slab Hole camp for the sand bar at the mouth of Lewis Creek: "Going out of Little Otter and rounding the willowy sand-point, the two craft fared across the bar toward the seining ground. Near them on the right curved the flat shore marked here by willows and farther on by a pale of rushes, the border of a great marsh that was walled south and east by the ancient forest, on the north by the great water maples and buttonwoods of Lewis Creek—a bay of rank marsh herbage with islands of button-bush dotting its fresh verdure with clumps of darker green."

That was the marsh of Robinson's youth, and it repeatedly provided the rich setting for his fictional work. But in Robinson's essays, the present intrudes, and the marsh becomes the focal point for his unrelenting lament about his era's blindness to environmental conservation. The following passage from an essay published in 1896 is worth quoting at length, not only because it voices Robinson's melancholy over what used to be, but because it is a vivid description of the marsh:

The yearly growth of lily-pads, wild rice, rushes and sedges, is the same that it was forty years ago, but I miss the old familiar trees that bent over the marshes from the shores that are now only naked banks of clay, and the broad primeval forests, in whose place are now only dreary acres of stumps and scant herbage. I miss the once teeming wild life of the marshes. I do not see one duck, nor hear one, and few bitterns, and only one heron; there are not so many kingfishers, and even the blackbirds are scarce, scant flocks of them rising in a scattered flutter out of the wild rice, where once arose a black cloud with a startling thunder of wings that made one's gun spring toward his shoulder in expectation of larger fowl worthier of its lead. Some alarmed fish break the water with retreating wakes at my approach, and I see some signs of muskrats, the floating remnant of their late suppers and early breakfasts, and hear sounds behind the green arras of rushes, splashes, plunges and smothered squeaks, that I attribute to these little representatives of their long-departed bigger brothers, the beavers. It is comforting to one who loves the inhabitants of the wild world to know that some of them still fairly hold a place in it in spite of all persecutions and all encroachment of civilization. Every spring three or four hundred or more of these fur-wearers are taken out of the marshes . . . by the trappers and shooters, and yet there are muskrats, and the chance of their continuance for years to come, for it is hardly probable that the water and the marshes will be improved off the face of the earth within the lives of several generations of men.

I notice as many as ever of the marsh wrens' nests on their supports of gathered rushes, and hear the rasping notes of these birds, always reminding me of those well-intentioned persons who have neither voice nor tune, but will always be trying to sing.

Button bushes are not worth cutting, even in malicious spite of woody growth, and their wide patches of scraggly, impenetrable tangle flourish and bear balls of purple buds, white inflorescence, and green and brown fruitage, whose bristling rotundity nothing seems to assail.

There is promise of a great crop of wild rice this year, but the old-time harvesters will not come in any force to gather it, as they did in the days of my youth. Then by the middle of September every stalk was stripped by the hordes of ducks, and the wet fields so cleanly gleaned by the throngs of blackbirds that it was a wonder how a kernel was left for next year's seeding. It is sad to think how the few survivors of that countless peaceful army will be harried by the more numerous army of gunners, and will not have a day's, hardly an hour's, truce given to them to rest and feed in the midst of this bounteous fare. Sometimes as one considers the ruthless bloodthirst of his kind, he is almost ashamed that he is of mankind, and then, considering how little better he is than the meanest of his fellows, and how much safer he is to be one of them than to be any wild thing, however harmless, he is humbly reconciled.

If Robinson's view of his fellow frequenters of the marsh seems nearly misanthropic, it is perhaps excusable as the inevitable outcome of his long and loving relationship with the marsh's myriad creatures. Sharing with fellow conservation-minded sportsmen of his era a concern for the health of wildlife populations, he went far beyond them in his ability to see animals and plants as knit together into a whole that, when broken, diminished humanity as much as it did the landscape. Robinson, without succumbing to the sentimentality of "nature fakery," empathized with individual animals he encountered. Occasionally he even took the point of view of the quarry in his more didactic writings about wildlife.

Nature's bounty for Robinson was such that even while aching with nostalgia for the more prodigious era of his youth, he could always find indications of hopeful continuity. Herbaceous plants frequently offered such solace. His portrait of the Lewis Creek marsh concludes:

> The blue spikes of pickerel weed bristle as of yore against the pale of rushes, and the white blossoms of sagittaria thrive there, above the spent arrows of their leaves, that some time since were shot up out of the mud and water by invisible sprites of the under world. The white dots that toss on my boat's wake as it stirs the border of rushes to a rustling of their intermingling tips I fancy at first are the breast feathers of some murdered waterfowl, or possibly a drift of castaway land blossoms; but upon examination they prove to be what my friend the botanist tells me is a species of buttercup—a milkman's buttercup it must be, so white and watery, yet nevertheless a pretty flower.

Robinson's natural history never aspired to science the way many of his contemporary nature writers—John Muir, John Burroughs, Bradford Torrey—did. The "milkman's buttercup" was white water-crowfoot *(Ranunculus tricophyllus or R. longirostris),* which, though a fairly rare species in Vermont, Robinson did not even bother to identify by its scientific name. In fact, this most astute naturalist of the Champlain Valley never troubled much with Latin names, and on more than one occasion he delighted in mildly ridiculing them. If he felt he needed the scientific name of a plant, he relied on his wife Anna, who would consult her *Wood's Botany* for identification. Once, when he brought home specimens of a marsh plant unknown to him, he sketched the plant in his notebook, noting next to it, "Arrow Head/*Sagittaria*—how I *do* hate these Latin names." Robinson even ignored the legitimate common names of plants and animals, preferring instead to call them by what names he and his neighbors knew them. Hepatica was "mouse-ears"; tupelo was "pepperidge"; sycamores were "buttonwood"; ruffed grouse were "partridge"; the white-throated sparrow was "whistling jack":

> The habit of using familiar names is hard to break in the ever-present temptation to make one's self easily understood. Ask the ordinary country boy whether there are any ruffed grouse in such a piece of woods, and if you get any answer but a blank stare it will be in the negative, possibly supplemented with the remark that he "never heard o' no sech critter." Meet him halfway and inquire for partridges, or come quite down to the level of his speech, beyond that unnecessary first "r," and he will tell you all he knows of those familiar woods-acquaintances of his.

Robinson's natural history was anecdotal and unsystematic, and he never attempted to portray his observations as objective, scientific data. They were the impressions of a man who had spent a lifetime hunting the woods of Lewis Creek and Little Otter, at first with a gun, and later without one. The authenticity of those impressions was validated again and again by knowledgeable readers who were not necessarily professional scientists. The reader with whom Robinson traded the most natural history (and ethnological) lore was Manly Hardy of Brewer, Maine. Like his father before him, Hardy had been a trapper and fur trader all his life, and knew the Maine woods as intimately as any white man. Somehow Hardy managed to keep abreast of the scientific literature, and might well have made his own contribution save for his natural reticence. (In a letter to Robinson he once remarked, "You ask why I do not write for *Forest and Stream*—The fact is I do not like to write anything where I have to speak of what I have personally done.") In Hardy's second letter to Robinson, on March 20, 1895, his critical comments about Henry Thoreau's *The Maine Woods* ("Thoreau saw but a very small part of our woods and did not know much about what he saw . . . [which] was about what one sees from a car window in going through snow sheds.") are followed by praise for Robinson's natural history: "Some things in your book surprise me. If you are not a regular hunter how did you get things so accurate? You have muskrat hunting exact in every point—but how did you know that coons in the spring visited each other? How that mink, especially males, did not take bait well in the spring? How that a vixen will drive a hound. And what puzzles me most of all how you could know just how an otter shot in the head would act. Now I have shot otters so and you describe their actions better than I can. I have reread your book lately and I cannot find a flaw."

It seemed as if Robinson could write over and over about the same animal, and never exhaust his store of knowledge. One of the first pieces he ever wrote (in 1875) for *Forest and Stream* was "Muskrat Trapping: Advice to a Beginner." He begins with tips on trapping along brooks, but then abandons the brook for more productive trapping ground—the marsh. "The sluggish stream moves with a scarcely perceptible flow between broad marshes, a rank growth of succulent rooted water-plants, arrowhead, pickerel weed, white and yellow water lilies, marsh marigold, sedges, and rushes of many kinds, furnishing a supply of food so endless that, with half a chance for their lives, the muskrats would fairly swarm in these marshes. As it is, in spite of the war raged upon them three-fourths of the year, there are enough left for a young trapper to learn on."

Robinson gives every detail about muskrat-trapping: how to set the tally sticks and make them easy to spot by slipping a strip of white birch bark in a cleft at the end; how St. Francis Indians and "Canucks" spear the muskrats in their houses in winter; which shrubs to use to make "bows" for stretching the skins. There is a great deal to learn about muskrat trapping in the Danvis tales as well. Antoine (whose forefathers first learned their techniques from the Indians) is as much a natural-born muskrat hunter as the mink, and shares his prowess with Sam Lovel. One afternoon he teaches Sam the muskrat's call, which Sam practices religiously that evening by their campfire. Sam becomes so proficient that his tutor calls to him from his bedroll: "Dah, Sam, you betteh stop you foolishin', fore fus' you know moosrat come an' bit aff you nose off."

Folklore crept in to Robinson's natural history accounts, but it did not always go unchallenged. The muskrat, Robinson said, "builds his sedge-thatched hut before the earliest frost falls upon the marshes. In its height, some find prophecy of high or low water, and in the thickness of its walls the forecast of a mild or severe winter, but the prophet himself is sometimes flooded out of his house, sometimes starved and frozen in it." It is not for lack of imagination that Robinson occasionally debunks folk beliefs; he was the most faithful folklorist who ever lived in the Champlain Valley. His own storytelling, both publicly via his writings and personally in his own family and community of Ferrisburg, enlarged the body of local folklore considerably. A revealing self-portrait occurs in a bit of reverie that Sam Lovel indulges in one afternoon at the Slab Hole Camp:

> As he looked eastward from the top of the bluff beyond the broad creek and above the wall of the woods, the first object that met his eye was Shellhouse

Mountain, and it struck him that the outline of its long crest, rising from the north end with one short curve and another longer one to the rounded highest point, thence sloping away to the south, greatly resembled a huge fish. Not far away a kingfisher hung steadfast for a moment on vibrant wings above the shallows, then dropping like a plummet, arose almost with the unbursting splash of his plunge, and presently proclaimed his good luck with a metallic clatter of his castanets. A fishhawk, cruising vigilantly above the channel, suddenly swooped and tore from the water a prize so heavy that, in labored retreat, he barely gained the cover of the woods in time to escape the swift onslaught of an eagle, lord paramount of all air, water, and earth hereabout.

"S'posin' you tackle Shellhaouse naow," Sam said as the baffled tyrant wheeled sullenly from pursuit, "I ha' no doubt you feel big enough t' think it wouldn't be more'n your sheer if 't was a fish."

Sam accepted these omens as auspicious of a good day's fishing, verifying what he had already felt in his bones, and was in haste to be off.

What Sam "felt in his bones" is what Robinson felt; they heeded the same omens. Daydreams were often as real as actual occurrences to both of them. There in his dreams Rowland Robinson was most like the Indians who visited the marsh. In an earlier day his daydream of the bald eagle seizing Shellhouse Mountain in its talons might have been transformed into a powerful myth told for generations. Shellhouse Mountain might have become *Namaswadzo*—"fish mountain"—if Robinson had been an Abenaki. Robinson's fascination for the Abenaki may have begun as a desire to record a vanishing way of life, but it quickly grew into an almost mystical kinship with them. They seemed, like him, to live equally in their imagination as in the world. The world was fantastically animate to the Abenaki that Robinson knew, and when they told him stories, he listened as a child might, not as a skeptic.

The liminal world of the marsh/swamp has its counterpart in the landscape of our own mind. Classic languages made this mind/landscape connection most directly; the Latin *limen* ("threshold") graded into the Greek word *leimon,* meaning meadow. Physiology recognizes the meadow habitat inside our mind by calling the center of emotion and imagination the limbic system, from the Latin *limbicus* for border or margin. It is the part of our minds where the fantastic flourishes, where everything is possible. For the Abenaki, the "edge" world, that of the wetlands, was the nursery for tales of the supernatural. One of the earliest stories a young Abenaki might hear would be that of *mskagademos,* the swamp creature, whose cry could often be heard coming out of the floodplain woods. The children were told that *mskagademos* was trying to lure them into the swamp where they would drown. There were also tales of *wiwiliamecq,* the great serpent, whose presence among the marsh plants was shown by places where mud and detritus boiled up from below. The log that a child might use as a bridge across a pool in the marsh or swamp might really be a *wiwiliamecq,* who would writhe furiously after being stepped on.

The Abenaki were not the only people whose imagination flourished in the swamp woods. The French-Canadians had their own menagerie of mythical beasts. There was the *loup garou,* a savage beast reminiscent of the Abenaki *gíwahkwa* (giant) and the Bastoniak's (the Abenaki term for whites—literally "Boston people") "Slipper Skins." There was also the French-Canadian will-o'-the-wisp—*le feu follet.* Robinson had ample opportunity to relate "Canuck" superstitions through Antoine; in *Uncle Lisha's Outing,* he relates a story about this phenomenon. One evening by their campfire, with the northern lights (the "Roarer Borer Alice," in Solon Briggs' scientific terminology) blazing to the north, Antoine sees a light moving about in the darkness of the marsh, and declares it to be *le feu follet.* Uncle Lisha and Joseph, disbelievers in someone else's superstitions, try to convince Antoine that the light is "nothin' but a jack o' lantern," but Antoine is not swayed:

> "Oh, no, no, no, dat ant be lamprin, sah; dat was feu follet! Ah do' know haow you call it in Angleesh, but he was very bat t'ing, Ah tol' you."
>
> "What is 't, Antwine?" Joseph inquired; "sort of a one-eyed lew grew critter sech as you was a-tellin' us on oncte?" . . .
>
> "No, seh, Zhozeff, he'll ant so hugly for keel someboddy lak de loup garou; he more kan o' funny for foolish someboddy. Ah'll had some experiments of it mahse'f, an' Ah'll goi' tol' you of it, me . . . one tam, w'en Ah'll han't more hol' as twenty year . . . Ah'll was rode 'long on mah leetly mare . . . an' Ah'll see lit over in mah fader hees farm where dey was be some swamp side of de meader. Ah'll ant know if he was somebody steal de hay or what he was do, but Ah'll t'ink he ant gat some beesiness dar, an' Ah'll go see what he was do . . ."
>
> "Ah'll go very softle as leetly maouses, but more furder Ah'll go de more furder de lit was go. Den Ah'll beegin for run fas', but he run more faster as Ah was, an' den Ah'll gat mad an' run more an' more faster, an' de more Ah run de more Ah'll get mad, an' de more Ah'll gat mad de more Ah'll run an' holler sco'ndrel name to it an' tol' it for stop, an' what beesiness he got, go to diab' for see his onc'—ev'ryt'ing Ah can t'ink, me, but he jus' jomp raoun' dis way, dat way, an de swamp an' say not'ing, only mek notion, an' dat mek me so mad Ah'll run on de swamp at it.

"Ah'll fregit Ah'll gat on all mah bes' clos'. Ah'll gat mah new moccasin, mah bes' tow trowser, mah han'some shirt mah mudder weave pruppus, an', seh, fus, Ah'll stubble mah toe an' sloop! Ah peek mahse'f careful, an' w'en Ah scoop de mud off mah heye, Ah'll see de lit go dance 'way off 'cross de swamp where somebody can' never go, an' den Ah'll know it was de feu follet."

Along with the Abenaki and French-Canadian spirits that dwelt in the swamp, there were also the creatures of the African-American imagination. Robinson heard all about them as an impressionable boy from Mingo Miles, who was brought as a slave from Rhode Island by Thomas Champlin to Vermont, where he became free at the age of majority. Mingo worked for Robinson's grandfather and was his first tutor in woods lore and folk tales. Mingo's monsters rounded out the multicultural mythology of the swamp woods for Robinson. His own appetite for myth was as strong as that of young Sammy Lovel, Sam Lovel's son, who used to beg "Mr. Tocksoose" for stories. In *Sam Lovel's Boy,* Sam Jr. and Uncle Lisha come upon a group of Abenaki engaged in basket making along a brook on the outskirts of Danvis. (Robinson himself visited just such a camp on Little Otter Creek in the 1850s.) Tocksoose, the leader of the group, acknowledges his feelings for Sam Lovel ("Lovett good man") and for his son, giving young Sammy a newly made ash basket for berry picking. Sammy returns often to the camp, and one day while hunting rabbits there with bow and arrow (purchased by Uncle Lisha for 25 cents), he asks Tocksoose how the rabbit got such long ears and hind legs. In the telling, Tocksoose also gives the explanation of the snowshoe hare's varying coat, which moves Sammy to ask how minks got black.

"You see, Wonakake—dat's otter—got mad 'cause mink ketch um so many fish, so he chase mink for kill it, an' mink pooty scare. He all white then jus' same weasel in winter, so otter can see it great way off; an' mink can't hide. So he run in where fire burn tree an' rub hese'f on burnt tree so he all black. Den he turn 'roun' an' walk back, an' by'm by meet otter run hard. Otter ain't know dat black feller, an' ask it, 'You see mink go dis way?' Mink say no, he ain't see it. Otter t'ink funny he can smell mink but can't see it, an' run on fast, but never ketch um mink. Mink like um color so well he always keep it, an ketch 'em more fish as ever, 'cause fish can't see um so easy, an' so he be black now."

The real "Mr. Tocksoose" was Joe Taxus, an Abenaki from St. Francis whom Robinson first met while ice fishing on Hawkins Bay one March day around 1880:

> The ideal angler has quiet ways, and observing that my third and last fellow fisherman—if I had a right to claim such fellowship—kept as steadfastly to his post as an Esquimaux to a seal-hole, never wasting a motion, I was attracted to him. He proved to be a Waubanakee of St. Francis, plying the gentle art here in the war-path of his ancestors. . . . [He] looked peaceable and kindly enough, but was at first as taciturn as his ancestors could have been, and as slow to be drawn into conversation as the fish to the companionship which I desired of them; but, baiting with tobacco and lunch, I at last drew some talk from him. He told me that he and a few of his people were wintering in a neighboring village, making baskets and bows and arrows. They found but little sale for these, and, for want of something better to do, he had come a-fishing. Years before I had known some of his people and through him I learned somewhat of my old acquaintances.

While talking with his new friend, the ice began "whooping like a legion of Indians," and Robinson feared it might "yawn beneath us and devour us." He asked the Indian what made the noise, to which he received the reply: "The ice." Not satisfied with so facile an explanation, Robinson coaxed Taxus for legends connected with the phenomenon of the booming lake ice, and got the following tale. (Note that the Abenaki "pidgin" is not rendered here as faithfully as in the passage from *Sam Lovel's Boy,* perhaps because as of that time—1883—he had not mastered it.)

> "You know that big rock in the lake off north—Rock Dunder, you call it? Wal, our people used to call that Wojahose,—that means 'the forbidder,'—cause every time our people pass by it in their canoes, if they didn't throw some tobacco or corn or something to it, the big devil that live in it wouldn't let 'em go far without a big storm come, and maybe drowned 'em. He forbid 'em. Wal, bimeby they got sick of it,—s'pose maybe they didn't always have much corn an' tobacco to throw 'way so,—and the priests all pray their god to make Wojahose keep still and not trouble 'em. After they prayed a long time, he promised 'em he'd keep Wojahose from hurtin' on 'em for a spell every year. So he froze the lake all over light every winter for two or three months, and then our people could go off huntin' and fightin' all over the lake without payin' Wojahose. That made him mad, an' every little while he'd go roarin' 'round under the ice, tryin' to get out. But he couldn't do much hurt, only once in a while git a man through a hole in the ice. That's the way I've heard some of our old men tell it, but I guess it's a story."

The marsh was a fitting place for Abenaki/white contact, for at the interface of land and water the two cultures were most alike. In the marsh the "Bastoniak" left behind most of his European technology. Steamboats and wagons had no place there; the *wígwaól* (birchbark canoe) and *maskwáolákw* (dugout) did. To successfully hunt in the marsh, whites had to adopt Abenaki ways; they took their cues from his reed duck decoys, his uncanny imitative calls, and blinds built of cattail and buttonbush. The two cultures—dominant and dominated—were equal there.

It was a harsh equality. The marsh was the last vestige of the ancestral Abenaki homeland to which they might come with impunity. They were not welcome anywhere else. Their great-grandfathers were once the lords of these estates of cattail and wild rice, equal with the otter and the osprey, but their sons and daughters were driven further and further north by white invaders, abandoning their last stronghold at Mississquoi just as the upstart nation won independence from Britain. Now they came sporadically from Odanak (St. Francis) on the Saint Francis River in Quebec, keeping to themselves as much as possible, except when they sold their wares—baskets, snowshoes, canoes—to the Bastoniaks.

Abenaki from St. Francis weren't the only outcasts who frequented the marsh and the adjacent swamp woods. There were other personages, whose names never appeared in Hemenway's *Gazetteer* nor on Beers's *Atlas,* who had made their homes there, if sometimes only transiently. John Cherbineau was one such resident, who lived just beyond the upper reaches of the swamp forest, west of the "path of boatless generations," as Robinson called the route that led from a landing on the East Slang to a favorite slough of the author's along the Lewis Creek floodplain. Sharing a love of the woods, Robinson and the old French-Canadian belonged to the same community, but Cherbineau and his wife were hardly full-fledged citizens of Ferrisburg. Along with inhabiting the marshy edge of the town, the Cherbineaus lived on the edge of the staid Protestant community's mores and manners. Though Robinson draws an endearing and sympathetic sketch of them, it was not a view shared by many of his neighbors, who leveled tremendous prejudice at French-Canadian immigrants:

> Stumps, young saplings, raspberry and blackberry briers held a far larger part of the deforested acres than did John's potato patch and cornfield, in the midst of which stood the little log cabin that, with its whitewashed walls and notched eaves, looked as little native to the soil as its tenants. I had not gone far toward it when a wide-brimmed straw hat appeared above the blackberry bushes, and as it moved slowly toward me in a halting, devious course, I discovered beneath it the broad, unctuous visage of John's "femme." Intent upon securing the last blackberries of the season, she was not aware of me till I called out to her, "Good morning, Marie. Where is John?"

My unexpected salutation did not startle her from giving chief attention to the heavily-laden bush before her, and her eyes and hands were busy with berries while she answered: "Good mawny! Mah man? Ah do' know 'f 'e ant peek hees onion. Ah do' know 'f 'e ant poun' baskeet, prob'ly. Yas, ah hear it," and listening, my ear caught the regular resonant strokes of splint pounding at the farther edge of the clearing.

Gathering and vending the various kinds of wild berries in their seasons, fishing and fish peddling, making baskets and braiding straw hats for the neighbors and storekeepers were the chief industries of this old couple, except when they once set forth on a grand begging tour, outfitted with horse and cart and a dolorous fiction of sickness and losses by fire. But they lacked one essential, a numerous, helpless progeny, through which to appeal to the benevolent public, for their own children were all grown up and scattered, and they could borrow but two of forty grandchildren, so the enterprise failed and they retired to private life.

"Lots of berries, aren't there?" I remarked, with a view to the old woman's encouragement.

"Oh, sang rouge; dey ant 'mos' any," she declared, in face of the evidence of laden bushes and a basket almost full of plump, dead ripe blackberries. "Dey ant honly few for beegin, an' dey all dry up 'cep' dees lee'l place!"

Robinson used the old Quebec couple as a model for Antoine's parents and the setting as a stage for a reunion of them with their son in *Sam Lovel's Camps*. Sam, along the same trail mentioned earlier, accosts an old man in "the baggy homespun trousers, red belt, and russet leather moccasins of his native land."

"Ah! Ha, ha, ha! you mek scare M'sieur! Bon jour, bon jour, M'sieur! You poot good, aujourd'hui M'sieur? Parlez-vous Francais, M'sieur? Non? Ha, ha, ha! me no parlez Anglais ver' good. Me come Canada las' printemps. Coupai le boi pour M'sieur Bartlette. Choppai de hwood. Onsten? Ha, ha, ha! Gat petit maison là, leet' haouse," pointing backward along the path and then beating his breast rapidly, "Jean Bassette, me. Me, ma femme, all lone, 'lone. Got garçon, boy, come here long tam, me can' fan', me sorry, oh! sorry, sorry. You no see it, prob'ly, M'sieur!"

Years later Sam and Antoine (in *Uncle Lisha's Outing,* 1897) revisit the "leet' haouse" to see if Antoine still has relations living there, and find smoke coming from the chimney and the sounds of a jigging fiddle and feet from the open door.

"Bah gosh, de smell an' de nowse was kan' o' Frenchy, don't it!" Antoine remarked as they drew nearer; but he started backward with an exclamation of astonishment when, still unperceived by the inmates, he cautiously peered

> in at the door. "Oh dey was too da' ks color mos' for mah rellashin," he whispered, as he fell back to Sam's side, "Dey was nigger!"

The two black men are the third class of marginal folk who find refuge in the swamp woods. The fiddler is "Jim," a freedman who gets his living from the woods and marsh. The dancer, "Bob," is an unfamiliar face, and it is obvious to Sam by his frightened demeanor that he is a runaway slave, being harbored here as distant from civilization as one could be and still be in Ferrisburg. Again, Robinson modeled his characters on real people he knew: "Jim" may have been inspired by Louis Dalton, a black fiddler of Robinson's grandfather's generation, or more likely by Mingo Miles. Jim's wife Nancy was a "handsome mulatto," perhaps borrowed from the nameless mulatto wife of old John Myers, a Dutchman (yet another second-class citizen of nineteenth-century New England) living in Robinson's youth near the west bank of Little Otter. Some indication of the status African-Americans had in Ferrisburg is hinted at in an essay written by Robinson in the 1890s ("The Path of Boatless Generations"):

> I stumble over the grass-grown foundations of the old Dutchman's house, and wonder to what quarters of the world was scattered the dusky brood that he and his mulatto wife reared here in the shadows of the locusts that he planted. There is something pathetic in the thought of those children, whose lot was cast with the despised race of the mother, though more of white than of negro blood ran in their veins. I remember one of them, a comely, sad-faced woman, harbored in middle age in the family of a negro, whom in her childhood she was too proud to marry. Poor Chloe, on what shore, far from this quiet stream you first beheld, were you stranded by the tide of years?

Another of Myers's daughters was courted by Mingo, and though he was unsuccessful, he "found consolation in the deeper-dyed charms of one Betsey, who held stern sway over his home till in a fit of revenge she burned the barns of Dr. Jonathan Cram, for which crime she was sentenced to the State's prison, thereby freeing Mingo from his second term of servitude."

Mingo was the "good black angel" of Robinson's childhood. In those early years, the young Quaker believed Mingo to be the only black man in the world, there being no resident African-Americans in Ferrisburg, and when the first "dusky passenger" arrived at their house on his way to freedom in Canada, Rowland's sister hailed him as "another Mingo." Their father's activity in the Underground Railroad brought many more Mingos, so that as an adult Robinson had a considerable knowledge of and empathy for the black race. He once remarked that in his family, prejudice was second only to slave-owning as the greatest sin.

Via Robinson's fictional work, one can get a fairly complete portrait of attitudes toward slavery in antebellum Vermont. In *Uncle Lisha's Outing,* Sam uses his knowledge of the marsh and swamp woods to outwit a slave hunter and his ally, and arranges passage for Bob aboard a Canadian sloop. Sam seems closest to the average Vermonter when it came to African-Americans; he had his share of prejudice, but his belief in individual liberty and his lack of economic reliance on slavery freed him to act as emancipator rather than oppressor. In *Out of Bondage,* Robert Ransom, whose father is a "Presbyterian and Democrat, and very bitter against Friends and anti-slavery people," harbors a runaway slave, and in another story ("An Underground Railroad Passenger"), the son of a zealous abolitionist marvels to see his proslavery neighbor aid a fugitive slave. When Abraham Thorne, the Quaker abolitionist, thanks his reformed neighbor, he gets the wry reply: "Sho! Abr'am, don't you never say a word about it. I wouldn't for all the world have it get out 'at I harbored a runaway nigger. Why, they wouldn't never call on me agin to help ketch 'em." But these attitudes were toward "passengers," not residents, and once African-Americans decided to make their homes south of the Canadian border, they often faced a different set of attitudes.

Joe Taxus, John Cherbineau, and Mingo Miles, and their fictional counterparts, are liminal characters, seen from the corner of one's eye. Indeed, they were all but invisible to many of Robinson's contemporaries, whose daily rounds rarely brought them in contact with the people of the swamp woods. When they did come in contact it was often with mutual mistrust, born out of a simple fear of "the other." The dominant white culture wished only that the marginal folk would stay invisible; the people of the liminal culture did their best to remain so. The dual movement toward nonentity is well illustrated in Robinson's stories, which un-self-consciously reflect the times. John Wadso, perhaps the first Abenaki that Robinson came to know, was "Watson" to Robinson's fictional Scottish pioneer Donald McIntosh; "his tongue could get no nearer his name than this." A similar inability of the Anglo tongue to master alien names is recorded in "The Shag Back Panther," whose "habitant" character Théophile Dudelant is known to his Yankee neighbors as "Duffy Doodlelaw." Robinson notes that along with Théophile, his transplanted Canadian friends had become "Littles, Shorts, Stones, Rocks, Grigwires, Greenoughs, Loverns, and what not."

African-American identity too was erased; no name but Bob or Jim or Tom ever sufficed. The abbreviation of names was adaptive and convenient, but the reduction to monosyllables hinted at a more dangerous mental reductionism about race. Not only individual identities, but cultural identity, was obscured by the dominant culture. Abenaki, Sokoki, Penacook, Cowasuck, Missisquoi were all "St. Francis Indians" to most Vermonters. Even those whites like Rowland Robinson who were interested in the cultural identity of the native peoples were confounded by the thrust toward invisibility. Though Robinson documented the first inhabitants of Ferrisburg as "Zooquagese," he never dug deep enough to discover that these *ozokwakiak* (the orthography for the Sokoki's own name for themselves) were "the ones who broke up, broke away," a name that told much about the cultural history of the tribe.

The Abenaki, the African-American, and the French-Canadian immigrant all adopted Yankee ways as best they could. They learned the language, wore the clothing, Anglicized their names, but they still kept to the periphery, where they could retain some freedom, some identity. In the swamp only the well-adapted—the silver maple, green ash, and swamp white oak, the wood duck and raccoon—survived the annual inundation of spring floodwaters. Along with the marsh, it had provided a sanctuary where creatures and creativity have been able to persist once they were gone from the cleared and cultivated uplands. The osprey and otter had their last stands here and have lately reappeared; the long-nosed gar, a "living fossil," holds out here against time; and liminal people yet frequent the marsh and the swamp woods, knowing this Eden is still open to them.

CHAPTER 3

Up Lewis Creek

N A SATELLITE image of the Champlain Basin, taken in late September, Lewis Creek shows up as a thin brownish-red line trailing southwest from the Green Mountain Front toward the brown blotch that is the signature of the Lewis/Little Otter Creek marsh. Here and there there are similarly colored flecks in the Champlain Valley—most belonging to the Overthrust Hills (Snake Mountain, Mount Philo, Pease Mountain, etc.)—amid a patchwork of reds, pinks, and grey-green. The reddish returns of radiation are fields of clover and timothy in full growth, the pink ones feed and sweet corn mostly, while the metallic tones are fields of grain or grass that have already been harvested. Shorn of their green cover, the bare earth reflects little of the infrared radiation that the satellite image processor is designed to detect.

What sets off Lewis Creek and makes it discernible from 500 miles above the earth is its trees; while most of the level, fertile land of the surrounding Champlain Valley has been kept relentlessly clear of its native vegetation, Lewis Creek and the other rivers that meander across the valley into Lake Champlain have been spared total deforestation. Even along the stretches where pasture runs to the river's edge, there are usually trees enough—a single file of silver maple or an intermittent row of struggling elms—to define the course of the Creek. Exactly what type of trees grow on the banks are indiscernible at this scale, as they are at the scale of a typical aerial photograph, say 1:20,000, although a fall photo will pick out stands of white pine and hemlock from their deciduous neighbors.

To know Lewis Creek's woods, one has to walk them, or canoe them. From the seining ground to about the Ball farm the banks are nearly part of the adjacent swamp woods, but above there, it is a primarily terrestrial landscape, secure from the annual spring inundation except for peak surges after snowmelt and April (or sometimes earlier) showers. In that season the silver maple, elms, yellow birch, red ash, and occasional sycamore that crowd against the river's edge bear fresh scars from the battering rams of ice and flotsam, and below them, the banks have not begun their green explosion. This riparian plant assemblage, like other "episodical" formations—desert herbs, pond mud, and deltaic deposits, all equally punctuated by rapid and severe fluctuations in the local environment—are "now you see 'em, now you don't" communities, their appearance and disappearance a caprice of the land and its weather. Lewis Creek's riparian region actually encompasses a suite of environments—clay banks, sand banks, gravel bars, cobblestone shores, alluvial flats, wet meadows, rock ledges, and more. The plant communities associated with each habitat not only materialize and dematerialize annually, but migrate back and forth and up and down the river as its point bars grow, its banks erode, and new channels are opened up. Right up to the river's edge in the moist, newly uncovered silt there are the forget-me-nots (the native *Myosotis laxa* and the alien *M. scorpioides*), horsetails (*Equisetum* spp.), bedstraws (*Galium* spp.), and smartweeds (*Polygonum* spp.). Two species of smartweed—*Polygonum amphibium* and *P. coccineum*—are so amphibious that often from a single rootstock two distinct forms of the plant arise, one in the water and the other out. In seasons when the water level has recently risen or fallen, the plants take on a growth habit transitional between these forms.

From a canoeist's perspective the most noticeable plants along Lewis Creek are the red-osier dogwood *(Cornus stolonifera),* its burgundy stems forever marking the river's edge; various alders (*Alnus* spp.) and willows (*Salix* spp.); sweet viburnum or "nannyberry" *(Viburnum Lentago);* and the vining, twining knot of vegetation that covers them—wild morning glory *(Convolvulus sepium),* common dodder *(Cuscuta gronovii),* purple clematis *(Clematis verticillaris),* wild cucumber *(Echinocystis lobata),* river grape *(Vitis riparia),* and greenbriar *(Smilax herbacea).* If one runs the Creek early one sees, back a bit from the river, ostrich fern *(Matteuceia pennsylvanicum)* pushing up through the grey silt, its uncoiling "fiddleheads" evoking images of *Maclurites.* With the ostrich fern are rich green patches of false hellebore *(Veratrum viride)* and false Solomon's seal *(Smilacina stellata).* A closer look would reveal sensitive fern *(Onoclea sensibilis),* golden Alexanders *(Zizia aquatica),* and a number of spring ephemerals—like hepatica, white trillium, trout lily, wood anemone—that wander toward the river from adjacent woods. Congeners of the wood anemone hug the river; river thimbleweed *(Anemone riparia)* prefers gravel banks, and meadow thimbleweed *(A. canadensis)* likes stony banks. In still rockier places, where Lewis Creek exposes bluffs of quartzite or dolomite, there are little rock gardens perched above the river, with early saxifrage *(Saxifraga virginiensis)*, columbine *(Aquilegia canadensis)*, and harebell *(Campanula rotundifolia)*.

As summer progresses, ranks of plants come into full bloom and then fade, so that the riparian swath has another dimension, a temporal one. In early June in some places it is all horseradish *(Armoracia lapathifolia)* or robin-plantain *(Erigeron pulchellus)*, while later the progression is seemingly all of composites—cockleburs *(Xanthium echinatum* and *X. speciosum)*, goldenrods *(Solidago flexicaulis* and *S. glutinosa* var. *racemosa)*, dwarf ragweed *(Senecio pauperculus* var. *Balsamitae)*, and tall coneflower *(Rudbeckia laciniata)*. Another composite, coltsfoot *(Tussilago farfara)* blooms at the opposite end of the season, its yellow flowers opening in early April and pointing themselves from the still cold ground toward the warming sun.

Some of the riparian plants are more elusive than others. Shortly after coltsfoot has blossomed, as its flowers turn to "water-balls," fluffy spheres of seeds ready to colonize new shores, a closely related plant once made its appearance on Lewis Creek. Sweet coltsfoot *(Petasites palmatus)* was collected from a cold, north-facing hemlock bank along Lewis Creek in Charlotte by Cyrus Pringle on May 29, 1892. A rare plant in Vermont, it has never been collected again in the Lewis Creek watershed, though botanists since Pringle have hunted it. The discovery was a fortuitous one, for Pringle's Vermont collecting activity had diminished considerably by that time. In 1876, the first year that Pringle actively collected, he made twenty-five botanical trips from his Charlotte farm; the following year his excursions had increased to fifty, and in 1878, he made sixty-two outings. Then in 1879, though he spent the entire month of August working in Maine and Quebec, he managed to botanize in Vermont on eighty-one days between April 11 and October 17. Between 1881, when he first went west to collect for Harvard botanist Asa Gray, and 1892, there had been little time to tramp the woods he had first known.

On most of his Vermont trips he would set out on foot from the East Charlotte homestead, usually alone but sometimes accompanied by his "cousin" Fred Horsford or his friend Dr. William Varney. The day he got *Petasites* Pringle had invited Willard Eggleston, a young botanist from Rutland, to come along; Pringle had been home a few months from a six-month expedition to the Mexican states of San Luis Potosi, Jalisco, and Michoacan. Two days later he was off in Mexico again. Eggleston recorded their Lewis Creek jaunt in a talk he gave to the Vermont Botanical Club in 1912, the year after Pringle's death.

> Pringle was just recovering from an attack of lumbago. One who took the smashing two days tramp with him that I did then would never have suspected it, however . . . we must have walked twenty or twenty-five miles, collecting a large number of plants along the way. We went to many of his favorite localities about the home farm, then up Lewis Creek a number of miles, over through Monkton to Flurona [sic] Mountain Bog, then back to his home. The entrance to Flurona Mt. Cedar Swamp is through a long beaver meadow. We rolled up our trousers and waded in barefooted, at one point the water was thigh deep. We then wandered barefooted for four or five hours, through the extensive forest of old growth cedars. About dark we got back home for dinner, and if anyone was troubled with lumbago that night, it was not Pringle. The next morning we walked to Mt. Philo and visited several of his favorite places on the mountain. . . . I can remember but few of the plants Pringle called my attention to those days but the following are a few of them. The first one was *Botrichium simplex*. With this plant he had some fun with me. He showed me a cradle knoll upon which we found two specimens. I looked for as much as five minutes and gave it up, but they were there, nevertheless. Afterwards we found quite a patch of the *Botrichium* in another locality. We then saw *Aplectrum, Petasites palmata,* and the Lewis Creek station for *Pterospora andromeda*.

For Pringle, the Creek was a corridor into the world of plants, and its variety kept him challenged for many years. There was not just the riparian corridor to botanize: There were the cedar swamps and bogs of Monkton and Bristol; the "Red Sandstone flora" of Mounts Philo, Fuller, Florona, and scores of unnamed quartzite outcrops; the limestone headlands on Hawkins Bay and the Gardiner's Island refugium; old-growth pine woods on sandy terraces. The Lewis Creek watershed provided Pringle with that opportunity that he later encouraged fellow Vermont collectors to seek out: "Share the secret of success of an old collector—quit the broad plain of dull sameness, seek out every possible situation of exceptional character, and look to find amidst peculiar conditions rare and localized plants."

The Lewis Creek avenue extended into time as well as space, but it was the watershed's other lover, Rowland Robinson, who traveled that road. Unlike Pringle, he was not born in the watershed, but just south of it, in 1833 in the East Room of the house his grandfather built along the stage road, near the upper reach of East Slang. The barely perceptible divide between the Slang (properly a part of the Little Otter Creek watershed) and Lewis Creek lay less than a half-mile to the north. As a boy, Rowland ("Row" rhymes with "how," not "hoe") Robinson attended a one-room schoolhouse that stood on part of the Robinson farm, and then his schooling continued a few hundred yards away, at the Ferrisburg Academy, a two-story brick building that embodied the hopes of the Ferrisburg community for a better future for their children. Robinson called himself "an unwilling pupil," finding more learning in his father's library and in tramping the nearby fields and forests than he did at the academy. The prospects for the young man seemed fairly straightforward; he would farm his father's land, as a modern, educated farmer, open to new agricultural techniques and tools, unafraid to explore new markets for his produce should they arise, and increasing by his mind and body the prosperity of the little republic around him.

Robinson would have made a wonderful farmer: His orchard was one of the finest in the Champlain Valley; his vegetable garden produced many vegetables not commonly raised in the region; he was a gentle and patient shepherd; and in the churning room behind the house he made excellent butter. But on the covers of the wooden butter tubs, bound for the Burlington or Boston markets, he would make sketches and illustrate humorous dairy stories. On the paper wrapped around the butter he drew forest scenes. Those wrappers and covers, not their contents, marked his calling. Along with the penciled butter tubs, he expressed his visions on canvas—with pen and ink, wash, watercolor, and oil—and even on big bracket fungi, which he foraged from the locusts and elms in the yard (these he carved with scenes and turned them into block prints). He had inherited his mother's—Rachel Gilpin Robinson, a New York artist before she married Robinson's father—artistic temperament, and it took precedence over the prosaic pursuits of farming. In 1858, he set out for New York City to try his fortune as an artist. There he found a job as a draftsman's assistant, and eventually sold drawings to *Frank Leslie's Magazine,* the *American Agriculturist, Rural New Yorker, Hearth and Home,* and other popular periodicals. Though he became discouraged after a while and returned to the farm, he went back to New York in 1866, working as a draftsman for Orange Judd's publications, and this time he found a wider audience for his artwork in *Harper's Bazaar* and *The New York Weekly*. They were mostly cartoons, in their themes typical of the times, and of Robinson himself in the quirky, idiosyncratic way he portrayed people's dress and speech.

In 1870, Robinson married Anna Stevens of East Montpelier, an artistic, literary woman whose gifted family had been friends of the Robinson's for generations. He continued to work for stretches in New York, but in 1873, the work had become a strain to him, and he returned for good to the Ferrisburg farm. His vocation became relegated to his sketchbooks and diaries, until 1875, when Robinson provided 15 illustrations for Anna's article "American Game Birds," which appeared in Moore's *Rural New Yorker*. That same year, he began writing for *Forest and Stream*—mostly descriptions of trapping muskrat and mink, and like subjects.

In his writings as well as his drawings there was an unmistakable sense of play that endeared Robinson's work to his readers. He had come by that playfulness primarily as a fox hunter. "Play" was the term Robinson and his fellow hunters gave to the fox's habit of making small circles in the woods when retreating from the hunter and his hounds. The Yankee's method of fox hunting was founded on this habit:

> If he "plays" in small circles, encompassing an acre or so, as he often will for half an hour at a time before a slow dog, you cautiously work up to leeward of him and try your chances for a shot. If he encircles the whole hill or crosses from hill to hill, there are certain points which every fox, whether stranger or to this particular woodland born, is *likely* to take in his way, but not *sure* to do so. Having learned these points by hearsay or experience, you take your post at the nearest or likeliest one, and between hope and fear await your opportunity.

Though he had hunted foxes all over his part of the valley, from Mount Philo south to Vergennes and Hawkins Bay east to Monkton Ridge, his favorite spot was a little notch in Shellhouse Mountain, due east of the Robinson farm:

> Such a place is this Notch, toward which with hasty steps and beating heart you take your way. When the fox returns, if he crosses to the south hill, he will come down that depression between the ledges which you face; then cross the brook and come straight in front of you, toward the wood-road in which you stand, or else turn off to cross the road and go up that easy slope to the south hill, or turn to the left and cross on the other hand. Standing midway between these points, either is a long gun-shot off, but it is the best place to post yourself; so here take breath and steady your nerves.

Posted there, intent on the possibility of a shot at the passing fox, Robinson entered a state of heightened consciousness peculiar to the poised hunter. He heard the trickle of the brook, the "penny trumpet" of the nuthatch, a woodpecker's hammering; he saw the hues of the landscape around him—gray tree trunks, blue-gray sky, flecked with the russet and gold of overhanging leaves, and "frost-blackened beech drops . . . the dull azure berries of the blue cohosh . . . milk white ones, crimson-stemmed of the white cohosh, scarlet clusters of wild turnip berries, pale asters and slender goldenrod. . . . His body is all sentient in its singleness of purpose—to kill 'Reynard.'" He gets his chance, fires, and then rejoices when the smoke clears to see the fox "done to death, despite his speed and cunning." In fox hunting, Robinson recapitulated the history of life, the unrelenting pursuit of predator after prey. It was the same epochal dance that produced the fingers clutching the gunstock, the eyes that discriminated the red spot that grew out of the russet leaves, and the brain that played and simultaneously recorded its play in words. That brain was at its best when honed on something wild, a fleeing fox or duck or the flashing scales of a brook trout. There, far from man, he was humbled and most human.

Robinson's playfulness eventually found expression in his writing via fiction. In 1881, "Uncle Lisha's Spring Gun," the first tale of the "Danvis folk," appeared in *Forest and Stream* magazine. It was a classic tale of human folly, Uncle Lisha and his kin and neighbors believing him on the point of death after an encounter with a bear, only to have the rug pulled out from under themselves when the local doctor arrives to declare that the guts the old man has been clutching against his belly are not his own, but the bear's! After that incident, Uncle Lisha Pegg's shoemaking shop became a place for the Danvis men to sit and "swap lies," making it the focal point for two decades of Robinson's fiction, which was published episodically in *Forest and Stream,* then gathered into book form. The tales are filled with Robinson's playful, skeptical appreciation of the foolishness of his fellow man.

In dethroning humans Robinson used an ancient storytelling device—the trickster. He'd grown up with Mingo's Br'er Rabbit tales, Yankee tales of the clever fox, and even heard Abenaki stories of *Azeban,* Raccoon. There was no single archetypal trickster in Robinson's stories, but an ever-changing array of characters who one moment might be the perpetrators, the next moment the victims. Antoine Bassette became the trickster in a wonderful Robinson retelling of the "stone soup" folktale in *Uncle Lisha's Outing.* Near the bridge over the East Slang, Antoine catches and kills a huge snapping turtle, called affectionately by him "Onc' Mud Turkey." Continuing his walk, he comes upon a pauper woman's cabin on the edge of the swamp woods. The woman whines that she's had nothing "but pertaters an' johnny cakes an' green corn t' eat for a fortnight," and Antoine wonders aloud why not, with duck and fish so plentiful nearby. She responds that they have no gun with which to hunt, and so Antoine suggests they eat "mud turkey." Disgusted, the old woman says she'd rather eat snake, so Antoine figures out another way to provide for her. He offers her some of the "chicken" he is bringing back to camp, and hands over a choice portion of freshly dressed turtle meat. He is on his way before he gets a chance to revel in the success of his trickery, but gets another opportunity back at camp. Though delighted at the prospect of fresh meat as a break from their steady diet of duck, his hunting comrades disdain the turtle. When Antoine leaves his kettle of turtle soup for a few minutes, Uncle Lisha and Joseph dump out the soup and substitute duck for the turtle. As they settle down to await Antoine's return, they hear him cry in distress from below the cliff at the Slab Hole. Not finding him, they return to the campfire, where they find Antoine. The three men then set to their supper, Antoine with unquestioning faith and great appetite, the two conspirators in mock tentativeness. When Antoine asks how they like it, Joseph says "it r'ally don't deem 's 'ough it was a turrible sight diff'ent f'm duck," and then confesses their trickery to Antoine. Antoine in turn tells them how he accidently discovered what they were up to, then drew them away from the pot so that he might restore its contents. Reaching into the kettle, he pulls out a cedar chip which he'd earlier placed there in case he needed confirmation of his tale. Knowing what they have eaten, the duped men begin to feel sick, whereupon Antoine ribs them: "Ah'll look see if de moss beegin for grow on you back, Onc' Lasha. Oh, don't you go crawl on de ma'sh."

Immune to this sort of deception, Sam wanders into camp, sees the empty turtle shell, and asks for some "mud turkle," confessing that he has always wanted to try it. Not only was he never fooled, Sam, more frequently than his comrades, assumed the role of Hermes the trickster-thief. In the beginning of *Sam Lovel's Camps,* a contrary local tries to oust Sam from trapping on the Little Otter marsh, saying that the muskrats are his. A little later the fellow appears again, poling a skiff amongst the buttonbushes, and calls out to Sam for a "chaw o' gum." Sam tells him he has one, but instead of a plug of tobacco he delivers a "sounding fisticuff full in the face." Elsewhere in the same book Sam comes upon an angler who has caught a dozen good-sized bass upstream from the mouth of Lewis Creek. The proud fisherman shows Sam his method, which to Sam's horror is simply a matter of catching the fish as they are about to spawn in their sandy beds on the bottom of the Creek. Offering Sam the chance to do likewise, the lazy and destructive fisherman is dumbfounded when Sam catches a good three-pounder and then promptly returns her unharmed to the water. "I allers thought it was a pleggid mean trick to ketch traout on th' beds, an' I guess this hain't no better," Sam scolds.

In these situations Antoine and Sam are a white man's Coyote, a Yankee's Raven. Their antics rest alongside a ten-thousand-year-old tradition that encompasses Coyote, Raven, Rabbit, Raccoon, Spider, Bluejay, Mink, Gluskap, Azeban, and Hermes. But Robinson's "trickster" anecdotes were as much morality plays as they were expressions of the universal polar structure of the psyche. The particular moral code that Robinson aimed at with these tales was one of proper stewardship of natural resources. The trickery is always leveled at those who squander and overexploit wildlife, or attempt to monopolize it. "Walk humbly and gently over the earth" is the moral lesson in each of them.

Trickster tales were not the only vehicle Robinson used to voice his particular land ethic. Like other writer/naturalists of his era, Rowland Robinson was moved by his intimate relationship with nature to speak out explicitly and directly as preservationist and conservationist. This voice comes through repeatedly in Robinson's writing, and was at the time perhaps the only Vermont voice consistently raised against the development then being visited on the Champlain Valley. Writing fictionally about a time already a generation past, he often made it his measure of the contemporary state of things. This was Rowland Robinson's lament, that the world had changed somehow from the time of his father, not only in the human communities that were the fabric of life for most, but in the natural communities which were just then coming to be seen as a vital part of that fabric.

That lament sounded strongest when Robinson wrote of his "three rivers": *Sungahneetook,* the Fish Weir River of the Abenaki, which Robinson wished had never been degraded by the French to its present meaningless name—Lewis Creek; *Won a kake tuk-ese,* Little Otter Creek; and *Petonktuk,* or the Crooked River, known to us as Otter Creek. The identification with aboriginal names went deep with Robinson, and like other nature writers of the era, he had even at an earlier time had an "Indian" pen name—*Awahsoose,* Abenaki for "bear." Robinson's knowledge of these watercourses was more like an Abenaki's than a European's—he could detail the scenery and events at almost any spot along their length. They were for

Sketch by Rowland Robinson of his skiff along the bank of Lewis Creek. Courtesy Rokeby Museum, Ferrisburg, Vermont.

him thoroughfares back to a more primeval time, and when the trappings of civilization pushed toward them, he cried out passionately. At the close of *Along Three Rivers,* a wonderful description of Robinson's timeless watery paths, his lament was never stronger:

> Our three rivers, shrunken though they be from their old estate, flow on over their ancient beds. . . . Within my memory, the shores of these beautiful streams were clad with noble forest trees, but year after year the work of their destruction has gone on till scarcely one of the original remains, and the miserable havoc still continues, extending even to the humblest shrubs, and is like to continue, though the mischief wrought in the increased evaporation of the streams and the wearing of the banks must be apparent to those who regard nothing in nature, but with a view to its utility. . . . When the work of the spoiler is completed and butchers who miscall themselves sportsmen have killed the last heron, bittern and kingfisher that still give a touch of wild life to scenes already grown too tame, our once beautiful rivers will have few attractions for those who seek the finest gifts the country can bestow upon its real lovers.

In much of his fictional work, Robinson weaves into his narratives insightful commentary on conservation practices, or more exactly, the lack of such practices. In *Sam Lovel's Camps,* published in 1889, Robinson levels his aim at a variety of despoilers. In a chapter that romantically describes Sam, Pelatiah Gove, and Antoine's excursion on South Slang to shoot pickerel, Robinson writes:

> Guns were booming all along the shores—the thin report of rifles spitting out their light charges, the bellow of muskets belching out their four fingers

> of powder, tow wads, and "double B's," and giving one's shoulder a sympathetic twinge as he thought how the shooter's must be aching—all proclaimed that it was a sad day for the pickerel that had come on to Little Otter's marshes to spawn. Probably not one man of the fifty who were hunting them there had a thought of what the fish were there for, or would have cared if he had. There were too many pickerel, and always would be. There could be no exhaustion of the supply of them nor of any other fish. Any proposition to protect fish and game of any kind, to prescribe any method of taking, to limit the season of killing, would have been thought an attempt to introduce hated Old World laws and customs. Hunting and fishing were the privileges of every freeborn American, to use or abuse whenever, wherever, and however he was disposed. And he could not live long enough to see the end of it, for why should there not always be fish and game as innumerable in all these unnumbered acres of water and marsh and woods? Alas! why not?

Some justice is done at least—when Pelatiah blasts a big pickerel a few yards away, the recoil of his gun topples him into the marsh.

Robinson's disgust with bad fishing practices knew no bounds. As fond as he was of Antoine and his fellow "Canucks," he believed they were generally guilty of overfishing. Describing Antoine fishing for bullpout from the old rotted bridge over South Slang, he says:

> To the Canadian a bullpout was a bullpout, to be taken at any time, by any means, and without regard to its condition. If he ever thought, as doubtless he never did, how the continuation of his most prized fish depended on procreation, doubtless he would not care, for what Canuck ever did? Apparently it is their belief that fish were created solely for them, and belong to them alone, and that they have a right to take them in any manner, as they will if they can, the last one to-day, though there should be no fish for any one forever after.

As avid a hunter as he was an angler, Robinson was equally sensitive to conservation applied to furred and feathered creatures as he was to finned. In *Hunting Without a Gun* he often cautions against overhunting, most poignantly when talking of Jigwallick, the great marshy indentation of Lake Champlain between Lewis and Little Otter Creeks. Once it was a favorite haunt for wood duck, and he found it deserted by them due to the incessant shooting of duck hunters. All this is not to say that Robinson was against hunting. His stories, despite their criticism of poor conservation practices, confirmed the basic value of hunting and fishing. In the early essay on fox hunting, Robinson sheds no tear for the dead fox:

> The shade of sadness for a moment indulged over the vigorous life so suddenly ended by your shot is but a passing cloud on the serene happiness you feel at having acquitted yourself so well. If you had missed him, it would have been but small consolation to think the fox was safe. The hounds having had their just dues in mouthing and shaking, you strip off Reynard's furry coat—for if English lords may, without disgrace, sell the game they kill in their battues, surely a humble Yankee fox-hunter may save and sell the pelt of his fox without incurring the stigma of "pot-hunter." At least he may bear home the brush with skin attached, as a trophy.

Once when Sam brings a partridge he has shot to Joel Bartlett's, he is met at the door by Bartlett's wife, who says, "I can't understand how people can enjoy killing things, such pretty things as partridges." Sam's reply: "They hain't no prettier 'n posies, an' it kills posies tu pick 'em. But that hain't what you pick 'em for. It's to hev 'em." When Mrs. Bartlett protests that picking doesn't hurt the posies, Sam responds with sly intuition: "That's more 'n we know, bein' 'at we hain't posies."

All these fictional anecdotes reflect the nonfictional sentiments of an avid conservationist—along with satirizing poachers and slob hunters in his "Danvis tales," Robinson made some very real life efforts at game preservation. The most consistent and perhaps effective outlet for raising the alarm against environmental degradation was through letters to *Forest and Stream,* which was the first conservationist periodical in the United States. In its first issue (September 1873), editor Charles Hallock printed articles entitled "Man the Destroyer" and "The Balance of Nature." In subsequent issues he drew attention to the imminent extinction of the American bison. Ten years before the Audubon Society was formed, Hallock acquainted American readers with the details of the London feather trade and warned of the consequences for American birds. When George Bird Grinnell—author, naturalist, and ethnologist—succeeded Hallock as editor in 1876, he continued the tirade against market and plume hunting, but extended his censures to sportsmen as well, urging them to limit their bags of game. The sensitivity to wildlife and reaffirmation of the ethics of sportsmanship contained in the columns of *Forest and Stream* was primarily produced by a growing vision of scarcity and a fear of extermination, as preached by Frank Forester and others.

Robinson's own stance was partially rooted in the scarcity scenario; his allegiance was to *game* protection. When the Vermont legislature passed a law in 1879 prohibiting the use of dogs in hunting grouse, Robinson wrote to *Forest and Stream* satirically suggesting that a law be passed prohibiting the use of guns in hunting, so that accidents might be reduced. That same year, when the season for woodcock was established (permitting hunting after September 1), Robinson wrote to say that in his experience in nine out of ten years all woodcock had left Vermont by that date. Like other scarcity proponents, Robinson was all in favor of the protection and restoration efforts that were just then coming into vogue. Of the poorly known and only locally renowned fishery called Lewis Creek, Robinson became the principal advocate and protector. Through the years he often acted as watchdog over the harmful fishing practices of many of his neighbors, who continued to use nets, shoot pickerel, and set night lines in defiance of the law. He was no self-appointed vigilante; the authorities asked for his help. In 1892, when C. A. Leonard of Hinesburg applied to Fish and Game Commissioner John Wheelock Titcomb for trout fry to place in Lewis Creek, Titcomb sought out Robinson for his opinion of Leonard. Robinson's political power in fish and game matters derived from his knowledge and his character, not from social and political connections. In 1901, the Vermont Fish and Game League passed a resolution mourning Robinson's death:

> As a man he was kind, generous, unostentatious. He was absolutely honest. . . . As a sportsman he was intensely interested in everything related to fish and game. He studied their habits and came to know of them as few do know. He approved of wholesome and wise legislation for their protection, and believed in an absolute compliance with these laws when enacted. He rated as selfish and mean any willful violation of them and wanted to see the offender brought to justice, and often with his own pen added to his discomfort after the law was done with him.

When the Vermont Fish and Game League was organized and incorporated in 1890, Robinson was the only one of its nine honorary members from Vermont. The others were influential members of the upper class from lower New England who traveled to Vermont—to the Batten Kill for trout fishing, in particular—for recreation. Many of the conservation movement's leaders were political and economic heavyweights—Teddy Roosevelt, Gifford Pinchot, Madison Grant, Owen Wister, Caspar Whitney, and Robinson's associate at *Forest and Stream,* George Bird Grinnell. These men, who through groups like the Boone and Crockett Club wielded tremendous power at the national political level, shared a common vision of wildlife scarcity. The first federal laws regarding wildlife reflect this vision.

In Vermont, the first fish and game code was enacted by the legislature in 1874, prompted by the rhetoric of sportsmen fearful of the diminution of their sport. Here, as elsewhere, there were a considerable number of elite members of the hunting and fishing fraternity; they had early on consolidated their political power in a group called the Vermont Association for the Protection of Fish and Game. W. B. Pettengill, Governor Horace Fairbanks, Henry Fairbanks, Judge Everts, E. J. Phelps, William Ripley, F. Stewart Stranahan, M. P. Gilman, Mason Colburn, and others were among its founding members. At a local level, Robinson allied himself with the Ferrisburg Sportsmen's Club, which he helped found in January of 1875.

The very first endeavor of the Ferrisburg Sportsmen's Club was to investigate the possibility of restoring salmon to Lewis Creek. If there was any single species responsible for the momentum which game protection efforts gained after mid century, it was the Atlantic salmon, *Salmo salar*. This king of the game fish once thrived in New England streams from the St. Croix to the Housatonic, but by 1850 had essentially disappeared from its native habitat west of the Penobscot River in Maine. In 1857, the governor of Vermont sponsored inquiries into the possibilities for artificial propagation of the Atlantic salmon and other fish, appointing George Perkins Marsh to the task. As a result of Marsh's report, two Fish Commissioners were appointed in 1867. The New Hampshire and Massachusetts Fish and Game departments also had their origins in the efforts to restore the salmon to its native waters.

Rowland Robinson was as determined as any salmon enthusiast in his quest for the holy grail of salmon restoration. Though born long after the salmon had departed Lewis Creek, he had grown up hearing stories of how huge salmon congregated at the pool below the dam at his grandfather's grist mill in North Ferrisburg. The farmers often speared a few ten-pounders while waiting to have their grain ground into flour. In February of 1875 Robinson wrote to Fish Commissioner M. C. Edmunds of Weston, Vermont, on behalf of the Ferrisburg Sportsmen's Club to see about introducing salmon fry into Lewis Creek. He made his case by stating the most important facts—the physical characteristics of the stream: "There is a portion of a small river near here some 3 miles in length from one mill dam to the next below. It is rapid, shallow, with occasional deep holes. The bottom is gravelly in some places, slaty in others. There are no fish in it but chub, dace, and such small fry. The banks are partially wooded. The water is pure and soft, but not very cold in summer."

Edmunds needed to see the stream before deciding on the potential for such an experiment. Lewis Creek was marginal habitat; he was aware that historically the salmon's southern limit on the Vermont side of Lake Champlain was Otter Creek, just two miles south of Lewis Creek. Upon seeing it, however, "he pronounced it as fine looking a salmon stream as he had seen in Maine or the British Provinces." On May 22 Edmunds brought 50,000 salmon fry from the hatchery at Charlestown, New Hampshire, and released them just above the lower falls of Lewis Creek. On June 29 Edmunds and Robinson walked the stretch between the lower falls and the second falls at North Ferrisburg, but after that Robinson and his compatriots from the Ferrisburg Sportsmen's Club were left to monitor the results of the introduction.

A year later Robinson wrote to Edmunds to inform him of the progress of the restoration effort. Along with his report, he urged Edmunds to promote the construction of fish ladders around dams on salmon streams. Unlike his grandfather, who simply never knew that the survival of the magnificent fish depended on adults ascending the Creek and smolts descending for their return to the sea, Robinson was well aware that any stocking programs would be thwarted if the fish were not allowed unimpeded passage along the stream. Edmunds replied:

> The necessity of fishways is a matter of the future but nevertheless seen and certain to become a necessity, and "necessity doeth all things swell," both as regards fishways and their builders.
>
> It affords a little pleasure to hear that you noticed a few of the young salmon we planted in Lewis Creek last season. I dare say they have disseminated themselves in the stream quite extensively above and below the point of their introduction. The fishing will have to be pretty thoroughly watched the coming summer, or many of your salmon will be captured as fingerling trout, and devoured by the idle fishermen along the stream. . . . The efficiency of Game Clubs to protect can be exemplified in the instance of affording protection to the young salmon you have in Lewis Creek, and I do hope it will warrant the expectancy of its friends.

As with all the early New England efforts to restore the Atlantic salmon, the experiment at Lewis Creek was doomed to failure due to the same factors that caused its decline in the first place. Dams at North Ferrisburg village and beyond at Scott Pond in Charlotte impeded the ascent of the fish to the breeding grounds, decreased the volume and flow, and consequently increased the temperature of the water. Pollution from woolen factories, sawmills, and tanning mills, as well as decomposed matter from cultivated lands, disturbed the meticulous habits of the fish, and sawdust accumulated on the spawning beds and clogged the migrating salmon's gills. Deforestation throughout the Lewis Creek watershed as well as denudation of the banks made the river entirely unsuitable for salmon: The volume of water was reduced and some tributaries dried up completely; water temperature rose and evaporation increased; sudden freshets, promoted by cleared uplands, disrupted natural logjams and removed the shaded haunts of fish; soil erosion cut away banks, changed the character of the bottom, and destroyed the salmon's spawning beds. Despite a heightened consciousness on the part of some fishermen, excessive and ill-timed fishing by improper methods contributed greatly to the depletion of the stock of fish. Seiners still cast their nets from the Hawkins Bay sandbar, and poachers took both young and adults whenever and wherever they could.

Had the salmon fishery at Lewis Creek lasted into Rowland Robinson's lifetime, there was more than pollution and overharvest to contend with. On the trip from the St. Lawrence the Lewis Creek salmon encountered clanging steamboats, mill machinery, blasting along the shore of Lake Champlain for building the railroad, and the rumble of the trains as they passed over the Creek on their way to Burlington. The advance of civilization was incompatible with the welfare of the salmon, which demanded the quiet and solitude of primitive conditions. It could not be civilized.

This and other nineteenth-century salmon restoration efforts also failed because at that time biologists believed all salmon to be alike. Only forty years ago did fisheries biologists learn that each one of the earth's Atlantic salmon rivers supports a strain of *Salmo salar* that is undyingly faithful to its particular riverine home. The stock of salmon that once populated Lewis Creek was distinct from the other tributaries of the Champlain watershed. When the salmon run in Lewis Creek was extinguished, it did not mean that the river was out of fish and needed to be restocked. It meant that a unique life form had vanished from the planet.

When exactly was the moment of extinction for the Lewis Creek subspecies of *Salmo salar*? As chronicler of the Creek, it was Robinson's role to discover this dismal date. His 1880 notebook records the information: "Stephen Hazard tells me again of the 'last salmon' of Lewis Creek. It must have been in the later part of June or first part of July, for they were going to haying when they found him—it was in the straight reach between the lower road bridge toward the lower end of it. This was the only one he ever knew of being caught. Thinks his weight 10 lbs." Ten years later the information appeared in *Along Three Rivers,* which was published in the Vergennes *Vermonter* in 1894: "Near the upper end of the loop, the last salmon known to have been taken in this river was speared with a pitchfork by Rowland, David and Stephen Hazard, three boys whose home was hard by, and the last of whom, born in 1800, died a few years ago. In the days of the generation preceding theirs, Lewis Creek swarmed with salmon in their season, and through their abundance the stream doubtless came by its Indian name."

The nineteenth-century chronicling of the last rattlesnake, last wolf, last catamount, last eagle, last salmon is a strange, macabre tradition. In Vermont, the last catamount stands stuffed in the Vermont Historical Society Museum in Montpelier, and there is reliable documentation of the shooting of this animal. History pays no attention to the voices of wild things, until their voice is extinguished. To record with such drama and precision the killing of the last animal is fitting; somehow the extinction of creatures becomes an index of progress, a yardstick against which we calibrate our domination of wildlife. Probably the "last salmon" of other streams has never been documented with such precision as Rowland Robinson accomplished. The salmon is invisible to all but the fisherman and mill owners, and it passed up its liquid highways to spawn ferociously, but quietly. It never howled at night like the dire wolf, never cried like the catamount to curdle the blood of whole neighborhoods, never took any human lives with venomous fangs like the rattlesnake. The salmon was never a scourge to be eliminated, but a boon, a blessing. Yet it went the way of its noxious brethren all the same. Settlers longed for the days when their sheep would no longer fall prey to wolves or panthers, but who could ever have foreseen the day when the big salmon ceased to run? Did anyone fear it in 1800?

That Robinson should have sought out Stephen Hazard (or vice versa) to nail down the event of Lewis Creek's last salmon was not serendipity. If Robinson was anything, he was the annalist of things past, and things *passed.* The myriad elements of the landscape that each had wonderful, fascinating lives, lives with real histories, were his passion. He had to document the pitchfork spearing of the last salmon, for it was a major temporal landmark in the life of his beloved Sungahneetook. Salmon were the life blood of the Creek in Abenaki times; they coursed into the stream each April, out each November, pouring their blood into the water, and their milt into the gravel in the headwater streams. Their flesh sustained small villages of Abenaki, and their presence was equally heralded by herons, eagles, and mink.

The spot where the Hazard boys speared that salmon is today, as it was then, practically naked of trees, and the banks and bed are all of blue clay; the pure, gravel-bedded, wooded stream that Robinson promised Edmunds still lies half a mile upstream. It is there where the clay yields to ledge that some ancestral memory of the great salmon runs used to bring Abenaki from St. Francis on their own monumental migration to Lewis Creek. They made the 150-mile trek by canoe to the first falls of Lewis Creek even into the 1830s. Robinson may not have been present for the last salmon, but he made it to the falls in time to witness a glimpse of the aboriginal past:

> I remember visiting with my grandfather a camp of St. Francis Indians, in a piney hollow that divides the south bank. They could not have been attracted by fish or game, but rather by the pleasant seclusion of a wooded place so near the highway and by the beauty of the little valley it looks out on, where the bright stream babbles down stony shallows from one to another quiet reach and eddying pool, that double in wrinkled reflections, wading kine, bank, flower, and pine clad slope, and, farthest and nearest, the crest of Mount Philo, where in the old days the Waubanakee kindled the signal smokes and looked broad upon the mapped wilderness for sign of enemy or answering signal of friend.

The Abenaki that gathered in that piney hollow did not call the stream that sparkled below their camp "Lewis Creek." To them it was still *Sungahneetook,* the Fish Weir River. It was there at the first gentle falls that the river came by its name. The little boy who stopped there with his grandfather could not have known the name. He only knew that these people were different from him; their skin was dark, their movements like deer, and their language sounded like water upon stone. The memory he took from the encounter was probably intelligible only in the ash basket or toy bow and arrow that his grandfather bartered or bought from them.

The ancient aboriginal name for Lewis Creek entered Robinson's consciousness years later, when he assumed the role of local historian to gather material for Abby Hemenway's *Gazetteer.* In October of 1858 he met John Watso, "an intelligent Indian of St. Francis," who told him that "this country, the northern part of the Champlain Valley, was occupied by a nation called the Zoòqùakgèes, of which the St. Francis tribe is a remnant. Their last abiding place in Vermont was at Missisquoi Bay. They called Lake Champlain 'Pe-tou-bou-qu,' signifying 'the waters that lie between,' that is, between the countries east and west. . . . Lewis Creek was 'Soon`gahnee-tuk,' which means 'The Fishing Place,' or Fishing River." A year later, Robinson got a chance to refine his understanding of the Abenaki name for Lewis Creek: "In talking with some St. Francis Indians this fall, I find that some of their names as above given are incorrect. . . . 'Sungahnee' means a *fish weir:* Sungahneetuk (the Waubanakee name of Lewis Creek) is the 'River of Fish Weirs.'"

How fortunate that Robinson preserved the name, for it was the only time it was ever recorded. Its authenticity was verified by Dr. Gordon Day, the foremost authority on the Western Abenaki language. According to Dr. Day, "With a few exceptions, Robinson's place-names are quite intelligible and the representation is about as close as he could be expected to make in an English orthography." His rendering of Lewis Creek's Abenaki name is remarkably good; Day's impression is that "Sungahneetuk" (Robinson alternated his spellings between this, "Sungahnee-tuk," and "Sun gah-nee-took") suggests the Abenaki word *senigànitègw,* or "stone works river." *Senigàn* ("stone works"), says Day, was an accurate description of the permanent stones in a fish weir that was planned to be used year after year. But a more likely explanation for Day is that Robinson's name represented his hearing of *kwsé nagan,* "fish weir." Though Day never obtained the word in a St. Francis dialect, he notes that ethnologist Frank Speck got its Penobscot analogue, *kwsé nagan,* in the 1930s.

As with all acts of naming, by speaking "Sungahnetook" Robinson claimed the river and made it his own. He often wished that "it had never been degraded by the substitution of its present meaningless name for the old significant one." But even this weak place name, an Anglicized commemorative name for a far away and forgotten French king, preserves history in its syllables. If one roots around a bit in the history of language, the toponymn that so displeased Robinson becomes more acceptable. The word "creek" comes down to us by way of the English, the words *crike* and *creke* forged at a time when the New World possessions of the English kings gradually slipped through their bent and benevolent hands. Their *creke* came perhaps from the Norse, who at the end of the first millennium A.D. saw new waterways as they raided across the North Atlantic. The Vikings in turn had borrowed from the Old English language: *krokr,* their word for hook, was akin to the Old English word *cradol,* in essence "cradle."

With "cradle," the seemingly understated and insufficient epithet "creek" comes full circle, encompassing through its etymology the idea of sheltering, nurturing, rearing. The creek becomes cradle; it is a place of origin, of birth. Lewis Creek has cradled them all: the Paleo-Indian ghosts who hunted mammoths in spruce–fir forests and whose utterings of names for the Creek or any other place we can only imagine; their descendant men and women—the Abenaki—who came to call the Creek "Sungahneetook"; and the white-skinned vagabonds who now people the watershed, having come with far-flung words and tools to be cradled here. The word could be excavated even more, digging from it the Old High German *kratto* for basket. Deeper, at the wake of Paleolithic times, on the Indian subcontinent there was a related Sanskrit word, *grantha,* which meant "knot." The watery cradle has become a knot, with no apparent beginning or end, a tangled bond of union between people and place.

CHAPTER 4

Wild Apples

A railroad in Vermont was almost undreamed of then, and there was no shadow of coming destruction brooding over the peaceful woods and waters.
—Rowland Evans Robinson

ODAY, IF ONE canoes up Lewis Creek from the Myer's Landing, where Rowland Robinson once imagined Sam Lovel to have spirited away a "blackberry"—a runaway slave—to Canada aboard a ship carrying Champlain Valley apples, one greets the Rutland Railroad line with nostalgia. As the canoe nears the trestle, one leans forward in hope of crossing under as a locomotive roars by, its steel rail rhythms singing of a glorious past. The iron horse, even in modern livery, calls forth a lament for a simpler era, even if we know that that era marked the beginning of the end for the Creek's salmon. For Rowland Robinson and his fellow Ferrisburg citizens, the railroad was the great and terrible harbinger of the future, not the hallmark of the past. The railroad was a mixed blessing for Rowland Robinson and the rest of his neighbors, both in Ferrisburg and throughout Vermont. It brought promise, energy, excitement, and tidings of a bright and benevolent future. As a boy of sixteen, Robinson must have followed its advent with considerable interest. Years later he wrote of it in *Along Three Rivers:*

> When the final survey for the Rutland and Burlington railroad was completed, a path was opened for it . . . from the East Slang to Lewis Creek through the original forest which then overspread nearly all the tract between the two creeks westward to the lake and still clad the left bank of Lewis and the right bank of Little Otter in almost primeval wildness. Construction

> trains from Burlington reached Ferrisburgh in 1848 or 9, hauled by a little locomotive as old fashioned as the place it was named for, Nantucket. It was a pigmy compared with one of the modern freight engines, but was the wonder and admiration of most of us, old or young, and its engineer and fireman, heroic figures, too grand for common mortals to have speech with.
>
> In a year or so, passenger trains were running and soon became ordinary and familiar objects.

Though the passenger trains that ran over the Creek may have become "ordinary and familiar objects," they still brought with them the unfamiliar and extraordinary. Along with new goods from Boston to be sold in Burlington and Vergennes, the train brought people with different modes of dress, conversation, and often a vastly different world view from Robinson's Ferrisburg neighbors. The railroad engine itself brought a new barrage of sensations: speed—the locomotives could move at almost a mile a minute over some of the level Champlain Valley terrain; and noise—until the train's arrival, the loudest sound heard in those woods was thunder in an August storm or a big swamp oak felled by a logger's axe. The feeling of the trembling earth as the train sped by was new also, nothing like the passing of a team, or the way the ice on Hawkins Bay felt as it cracked and shifted under one's feet. Those ephemeral tremblings left peaceful, consoling wakes, while the steam engine's sound and fury left an irrevocable, mildly horrific rent in the air.

Robinson eventually became unequivocal in his dislike of the railroad. In his notebook in 1878, he notes one dismal effect of the railroad—deforestation: "How *railroads* hate trees, all cut along the line, so I hate RRs." His other principal objection was that the railroad brought "away" to Robinson's doorstep. He had tasted that "away" soon enough when he had gone to New York to try his fortune as an artist, and it taught him that he preferred the meandering tributaries of Lake Champlain to the ceaseless current of human activity in the city. As he settled beneath the "star he was born under" to chronicle the lifeways of his people, those ways were being changed, and in no small part by the railroad.

But like all innovations, the iron horse took its place in Robinson's vernacular landscape as undeniably as the bow and arrow, the turnpike, and the telegraph. It became another element of place, as vulgar and wonderful as the hills of the Red Sandrock Range or Town Meeting. Just how commonplace the railroad became in a short while is evidenced in the diaries of Joseph Rogers, whose land encompassed a portion of the south bank of

Lewis Creek below the Hollow. Robinson bought the diaries at an auction a few years after Rogers had died. Robinson said of Rogers that he

> was an eccentric bachelor, the son of as eccentric parents, who were among the early settlers of Ferrisburgh. "Bach'ler Josey," as he was called . . . was a close and curious observer of nature and naturally her lover as well, who preserved his woods so jealously that her other lovers could wish that he might have lived a thousand years to keep watch and ward over his woodlands. For several of his later years he was confined to the house, and during this time, he kept a diary wherein occur many quaint notes concerning weather and birds, as observed from his window.

The diaries were filled with the most wonderful commonplace:

> March 6, 1870. Near dark meet Reynolds the meat man for a few moments—he does not know why we have Sunday on the first day of the week, only by tradition and education and custom.
>
> March 26. . . Cool, good flax day . . .
>
> May 14. The King Bird and red Robin has [sic] come. Red cherry trees begin to show blossoms, I see. People have largely turned their cattle to range or look over their pasture.
>
> May 21. A Pair [sic] tree east of house in full bloom, apple trees begin to show blows, a Red Plum tree done blowing . . . Dandalion [sic] begin to get water balls.
>
> May 24. See a red squirrel about locust trees. Moses [Yott, his nurse] got some dandelion root for me to put in cider.
>
> May 30. Cherry trees and apples 'bout here out of blow. Cherries first.
>
> June 9. Man came along with fish to sell, price for bull pouts a cent a piece . . .
>
> June 19. Pick berries so had strawberry shortcake for dinner.
>
> June 29. Hear Frank Mills buried a boy today drowned in Lewis Creek.

Though immobilized, Joseph Rogers often surveyed the farthest points of the landscape in his daily observations:

> March 27, 1871. Clear most to mouth of Lewis and Little Otter Creeks.
>
> April 28. Lake water and creek look white.
>
> April 30. Foggy, rainy, hardly see school house. Wind flag show almost calm. Fog has gone back so as to see some of creek water.
>
> May 21. Woodchuck is looking about school house, perhaps to sweep out and arrange seats for open school Monday. Dandeliner show largely in road south of schoolhouse. All about Mud Swallows from . . . barn are largely engaged at breaking the Sabbath by carrying materials from trough down road to repair nests.
>
> October 29. Eve see [Split Rock] lighthouse for first, leaves so off trees west of house.

Inserted in the diary is a newspaper clipping:

Trains Leave Vergennes

Moving North		Moving South	
3:45 PM	Mail	9:11 AM	Mail
6:48 PM	Express & NY Mail	2:35 PM	Express
3:18 AM	Night Express	10:20 PM	Night Exp.
8:00 AM	Mixed	6:50 AM	Mondays
5:00 PM	Local Freight		

The train was another of those distant points seen from his window that brought the world inside. For "Bach'ler Josey" the train held a place very much like that of his locust trees, the "Mud" (cliff) swallows, or a field of flax. It was an everyday object, with rhythms that lent order to the world. In many of his diary entries, after commenting on the appearance of the air—"rather thick," "pretty clear," "dark day," "stormy," and so on—Rogers notes whether he can see the railroad, and what the smoke from the locomotives tells him of wind direction and velocity. The 26-ton engines that shook the earth and belched smoke and carried people across the Green Mountains to Boston were simply a convenient anemometer to a housebound old man. As surely as the appearance of the Split Rock light told him that winter was nigh, the railroad announced the time of day and the condition of the valley winds.

For Rogers's fellow Ferrisburg farmers, the railroad was more than this. It meant new markets for their crops. Wheat and flour were among the first freight hauled south by the trains, even as improved transportation to the west already was knelling the approaching death of the Champlain Valley grain industry. Apples still held their own, though. Apples from Robinson's orchard sold at Faneuil Hall Market in Boston. At North Ferrisburg depot, there were two cider-mills—Stephen Ball's and George Kimball's. In the 1880s, Kimball's mill employed four men who manufactured 1000 barrels of cider each fall. He and Ball also produced hundreds of barrels of cider vinegar, much of it for the Boston market. Not only did these mills provide an outlet for the many local orchards, but they spawned other manufacturing enterprises. At Kingman's barrel factory in Ferrisburg, three to five thousand barrels per year were built to ship locally produced cider and pork.

The cider mills are long gone now, but if you are canoeing down Lewis Creek, and pull out on the close-cropped grass of the north bank just before the railroad bridge, you will take your lunch or a brief rest under a little grove of apple trees. There are about two dozen trees in uneven rows, and their western margin is the steep railroad embankment. Rotted railroad ties and loose ballast mingle with spent apples at the foot of the bank. The eastern margin of the swath of fruit trees is an old cut bank of the Creek, and if you walk away from the Creek gradually uphill with the apple trees, you get a view not only of Mount Philo, Fuller Mountain, and Shellhouse Mountain, but of what Robinson described nearly a century ago as "a great loop of the stream, half-encompassing many acres of meadow and marsh, much beloved of muskrats." In winter, when the channel and floodplain are equally white, an even-topped meadow of reddish-brown buttonbush marks the old oxbow.

The little orchard trapped between the Creek, its vestigial landform, and the railroad, is itself a vestige, for if you avail yourself of its forgotten fruit you taste not MacIntosh or Delicious, but some unfamiliar variety. This coarse, acidic, yellow-skinned apple is an Oldenburg, and away from the Creek a little there are Yellow Transparent and Tetofsky as well. These are "Russian" apples, come from their Asian homeland to the shores of Lewis Creek. Russian apples came into favor in Vermont in the 1870s, as a response to the failure of many varieties that had been brought by settlers from milder southern New England climes. They were originally imported by enterprising farmers from Wisconsin in search of hardy trees, and the United States Department of Agriculture began the general importation of Russian scions in 1870 and planted them on the Department grounds in Washington. From there they were distributed all over the country, usually by Congressmen.

The hype surrounding the Russian apples was reminiscent of the "crab-apple craze" of the 1850s, when aggressive fruit agents convinced people in northern New England that they could never grow ordinary apples. They "gave away" trees at fifty to seventy-five cents apiece with the admonition that it might be the farmer's only chance to grow long-lived apple trees. In some neighborhoods, Transcendent, Hyslop, Soulard, and other crabs became the dominant fruit tree, but not all of the hustling John Chapman's promises were realized. With the Russian apples came promises too. Foremost was the promise of hardiness: The very names—Arctic, Tetofsky, Northern Star, Alexander, and so on—suggested it. The trees were rumored to be free from disease and to bear early and abundantly. The fruit was reputed to be large and finely colored. These expectations were met, but there were problems. Many of the varieties introduced were worthless.

Most ripened too early and were poor keepers, having been introduced from a zone of shorter seasons to one with a longer growing season. Many varieties dropped their fruit largely before it had matured, and the young growth on trees was extremely susceptible to "fire blight." The fruit of the Russian varieties was often very thin and the flesh was coarser grained, making them unpalatable to most New Englanders.

The hope for the transplanted Russian apples lay in hybridization, whereby the desirable characteristics of the Russian trees might be wed to those of native varieties. Charlotte farmer Cyrus Pringle had received scions of fifty varieties of Russian apples from Washington in February of 1873, and the following spring he wasted no time in cross-fertilizing Northern Spy with a number of Russian varieties. By winter the hybrid seeds were planted in cold frames, their pedigrees carefully recorded by Pringle. It was too early for him to report on results, but in January 1874, Pringle expressed great expectations: "From the numerous hybridizations between the common and the Siberian species now being effected in this country, I hope for a class of apples that will be an addition to our present stock of varieties. The improved Siberian possesses a brisk, spirited flavor—a character half wild . . . I anticipate seeing [this character] imparted in some degree to the tamer flavor of our old sorts, and then there is no question but greater hardiness will be secured (but not always, by any means) for good apples."

The fortunes of apples in Charlotte, Ferrisburg, and the other valley towns of the Lewis Creek watershed rose and fell like the Creek itself, commanding attention and admiration for a few surging moments, then receding into oblivion. Apple trees had been on the continent almost as long as the white man, reaching the Maine coast in 1604, Salem, Massachusetts, in 1628, and the Massachusetts Bay Colony in 1639. By the time the English settlers pushed north into Vermont from southern New England, orchards of Fameuse apples were already established at the French settlement at Chimney Point. In the early nineteenth century, orchards were on almost every farm, and each farmer put into his cellar plenty of fruit to last the winter along with ten to twenty barrels of cider. By 1830, the first fitful starts of westward expansion and an increased concentration on sheep production lessened the importance of apple orchards, and then the Temperance movement brought wholesale destruction of some, since the main use of apples was still for hard cider. With the 1850s crab-apple mania, orchards again reached flood stage, but as quickly dropped below banks.

During the era when Cyrus Pringle was experimenting with Russian apples, orcharding in Vermont had taken on a new flavor. With scientific, more specialized farming, many farmers neglected or even abandoned their orchards, while others concentrated on fruit production for their cash income. Old, unkempt orchards became a ubiquitous element of the Vermont landscape, though relatively less common in the Champlain Valley, whose more equable climate and better soils lent themselves to profitable fruit production. Many farmers faced the same situation as Pringle described in a brief note entitled "The Old vs. the New" in 1871 in the *Country Gentleman:*

> At the time of my earliest recollection, there were scattered over some twenty acres of the heart of this farm [the Pringle farm in East Charlotte] and on its very best land, an old orchard of several hundred trees. By far the greater part bore only "natural" fruit. When I brought up against it a dozen years ago, I pronounced it a nuisance. It proved an enduring one. Although the wind and the caterpillar were constantly decimating its ranks, its outposts had not been called in. Scattered wide, it everywhere impeded neat and thorough cultivation; its shade stamped out here and there broad patches in the corn, and its falling boughs and twigs innumerable provoked the wrath of the mower. Yet it endured and contested the ground by inches; for present necessity was always on hand to suggest that here and there a grafted bough or a whole tree bore fruit worth from three to five dollars per barrel. So for years it was permitted to occupy ground which we were impatient to reset to young orchards or to devote to nursery purposes, however occasionally a part was removed, and the soil thoroughly prepared by deep tillage for the desired use. But, during the past fall, so favorable for such operations, the order went forth, and the old orchard came out by the roots, only a few lone trees in out of the way places being left to fill up the gap between the old and the new.
>
> Such old orchards are to be seen on every hand. How long, think you, after they have passed the point where they cease to be profitable, in view of the fine land they occupy; will they be suffered to linger on in all angular deformity to the exclusion of a successor? Or, if removed, will their stumps be left to disfigure the field and interrupt the course of the plow, when two men in an hour can grub out a tree with the aid of a team hitched to its top, to wrench off its lower roots?

All the while that apple trees were spreading neatly over the landscape by man's agency, a subversive, seemingly haphazard migration of *Pyrus malus* was also occurring. Along hedgerows, in young deciduous woods, on roadsides, seeds of the domesticated tree found a footing for spontaneous, uncultivated growth. No farmers set these trees out nor came each year to prune them, and their fruit was most often gathered by cows, horses, robins, woodpeckers, grouse, mice, squirrels, foxes, and raccoons. These were "volunteers," domesticated plants escaped from the hands of man to a naturalized state. The wild progenitors of Vermont's apples grew in Europe and Asia in deciduous forests, mixed deciduous/pine forest, forest edges, river valleys, mountain slopes, gorges and scrub areas, and for two centuries now the volunteer apples of the Lewis Creek watershed have been "escaping" back to similar habitats.

If you get back in your canoe and proceed downstream, immediately after you cross under the railroad trestle, you'll see on the right bank one of these volunteer trees. It's an open-grown tree, so it looks full and squat like its neighbors on the other side of the embankment, but its trunk is not as sturdy, and its branches go right to the ground in places. A beaver has nearly ringed the trunk, yet the tenacious volunteer tree survives, and the scores of suckers it is putting out ensure that should the beaver finally fell it, or should it lose strength from the girdling and die, it will be replaced by the most vigorous of the shoots now coming up out of the grass and goldenrod. It is not alone, this fugitive apple tree. There are others, straight-trunked and tall with few lateral branches, growing just beyond in the floodplain woods. Follow the Creek upstream to its sources and the scene would be the same. On any mid-May walk in the watershed, from the shore of Hawkins Bay to the deep ravines of Hillsboro, one would be hard pressed not to find the white blossoms of volunteer apples an almost constant companion.

"Volunteer" apples haven't always been called that. Rowland Robinson, who orcharded twenty acres at Rokeby, spoke once of his orchard trees' "plebeian kindred, the 'common' or 'natural' apples." For Robinson and his generation, another natural philosopher—Henry David Thoreau—had immortalized the trees, as "wild apples." Indeed, on Wednesday afternoon, September 25, 1850, Thoreau and his friend William Ellery Channing had passed this very spot, on Thoreau's sole venture out of the United States, to Canada. A few miles back at Vergennes, the Concord philosopher had gotten his first view of Lake Champlain, which he deemed "impressive, but rather from association than from any peculiarity in the scenery." The "association" Thoreau spoke of was one that ran back thirty years for him via the study of its geography on maps, but more immediately that fall through extensive readings in Cartier's *Voyages de découverte au Canada,* Champlain's

Voyages de la Nouvelle France and *Voyages du Sieur de Champlain,* and Lescarbot's *Histoire de la Nouvelle France.* Like Robinson, Thoreau was steeped in history, and it was primarily through these historical readings that Thoreau became fascinated by the "early history of the North American continent"—the history of European contact with Indians. Also, like Robinson, Thoreau had an opportunity for a firsthand encounter with *living* Indians—a group who camped along the Concord River in the fall of 1850—that gave him a unique, particular approach to "the red face of man."

Just as his interest in Indians became more scholarly, more objective, and more outward directed, Thoreau moved in the opposite direction in his rediscovery within himself of the essential wildness that first brought him to his interest in Native Americans. His autumn walks brought him upon wild grasses, wild house cats, wild muskrats, and wild men, to a pursuit of what he called "the wild." That autumn brought him also to the wild/volunteer apple trees of New England. On the September afternoon that Thoreau rode the rails over Lewis Creek past the spot where today a beaver-gnawed wild apple tree stands, there were likely no wild apples in sight. The paucity of wild apples was superficially Thoreau's main motivation for writing his "Wild Apples" essay. He was worried that the Temperance movement and the practice of grafting for table apples marked the end of the wild, racy fruit of the naturalized apple trees. Not only were wild apples on the decline, lamented Thoreau, but no one had even named them. He pictured himself at the cider mill, christening the "hundred varieties which go to a single heap" there. Though he drew his names from his experience with wild apples in Concord, the imaginary catalog he compiled serves equally well the fencerow and old field apple trees of Lewis Creek. There was the Wood Apple *(Malus corvi cristati),* the Frozen-Thawed Apple *(Malus gelato-soluta),* the Green Apple *(Malus viridis),* the Hedge Apple *(Malus sepium),* and the Slug Apple *(Malus limacea).* Our tree here on the bank of Lewis Creek might well be a "Railroad-Apple, which perhaps came from a core thrown out of the cars." Throughout Concord, as throughout Ferrisburg, Charlotte, and the other towns of the watershed, one could always find "the Apple that grows in an old Cellar-Hole *(Malus cellaris),*" or perhaps the most difficult to find wild apple, "the Apple whose Fruit we tasted in our Youth; our Particular Apple, not to be found in any catalogue."

The opening statement of Thoreau's essay, "It is remarkable how closely the history of the apple-tree is connected with that of man," suggests that it is more as symbol than as simple object that he scrutinizes the apple. His discussion is not confined to wild apples, but contains discourse on the cultivated apple and the native North American crab apple, *Malus coronaria*. In his mind the three types of trees become the three separate "races" of Americans in whose hands lay the fate of the nation. The cultivated apple was the European settler of the East's civilized communities; the native crab apple equaled the aboriginal people of the continent; while the wild apple represented Thoreau and his fellow Americans who had let the landscape naturalize them. It was with the last that his loyalties lay. He loved to go through the old orchards of ungrafted apple trees, or even more, to chance upon some vigorous cliff-grown wild apple, "planted by birds or cows." These trees, while retaining an aboriginal wildness, emulated the white "man's independence and enterprise," and bore, like Thoreau, the noblest fruit. Thoreau said it plainly himself: "*Our* wild apple is wild only like myself, perchance, who belong not to the aboriginal race here, but have strayed into the woods from cultivated stock." Such thoroughly naturalized individuals—himself, and perhaps Walt Whitman—were his generation's prophets, and their prophecy was that "in Wildness is the preservation of the world," that humanity would fulfill their destiny not by exploiting the American landscape, but by letting the land take hold of them.

In Rowland Robinson, Lewis Creek had its own prophet, a lesser prophet than Concord's, but with a like vision and one of whom the land did take a firm hold. Thoreau, had he lived to know Robinson through his writings, would have classed Robinson as a wild apple. Robinson's stock was all cultivated, well bred, educated, and town reared, and Rowland was no exception, but he strayed over the pasture wall and into the woods early, forsaking the four walls of the Ferrisburg Academy for the Porterboro Woods whenever he could. There, on Shellhouse Mountain, and in whatever other wild places the thickly settled lands of his youth had to offer, he grew among birch and pine and oak, to bear wild, ungrafted fruit. Like Thoreau, he often questioned himself as to whether the native, aboriginal apple wasn't the finest, and the one that would ultimately survive. In *Uncle Lisha's Outing,* Sam and a prominent lawyer from Vergennes are duck hunting on South Slang when they come upon a camp of St. Francis Indians. They observe a bit of canoe building and arrow making and then resume their paddling through the wild rice marsh in search of ducks. Sam's companion asks him how he'd like the life of their "red brethren," and his reply encompasses Robinson's meditation on the subject:

Rowland Evans Robinson. Courtesy Rokeby Museum, Ferrisburg, Vermont.

"It'll du well 'nough for a play spell naow an' ag'in; but it hain't jest the sort o' life for a stiddy business, leastways not for white men. Oh, I d' know, if a man had n't nob'dy but himself and things had n't gone jest right with him, but not if the's anyb'dy 'at he cares for. I hev wished I was an Injin, but I don't naow. An' I've tried it tew, for a fortni't runnin', up t'other Slang. An' it beats all haow easy a man settles daown tu that way o' livin', an' I b'lieve a man 's consid'able like a tame fox—oncte he gits loose he gits wild ag'in mighty easy. I feel it a-comin' on every time I git int' the woods, some sight or some smell 'at you can't sca'cely see ner smell, a-wakin' up suthin' that's b'en asleep sence the Lord knows when. 'T wan't in my father, an' I do' know's it was in my gran'ther, only as he hed tu hunt some for a livin'. 'T aint no wonder 'at you can't tame an Injin so 't he'll stay tame, wi' a hundred generations o' wild blood a-r'arin up in him wus 'n we c'n guess. An' 't ain't none tew easy for us tu quit livin' that way arter bein' in 't a spell. Why, it's jelluk leavin' the hum 'at I was born in an' reared in, tu clear aout from a camp 'at I've stayed in a week, an' if I come acrost it arterwards it makes me feel sort o' lunsome."

Though it is Sam Lovel's drawl we hear, the inner voice is unmistakably Robinson's, balancing Yankee pragmatism with the undeniable urge to "go native." Sam and his creator seem to say that there is no going back, no promise for civilized humanity in adopting the ways of the Abenaki; but he cannot shake the " homesickness" for the camp and its fire. The camp-fire was to Robinson what the wild apple was to Thoreau, holding the promise of modern man's reentry into that state of wildness which millennia of civilized living had deadened:

> If "the open fire furnishes the room," the camp-fire does more for the camp. It is its life—a life that throbs out in every flare and flicker to enliven the surroundings, whether they be the trees of the forest, the expanse of prairie, shadowed only by clouds and night, or the barren stretch of sandy shore. Out of the encompassing gloom of all these, the camp-fire materializes figures as real to the eye as flesh and blood. It peoples the verge of darkness with grotesque forms, that leap and crouch and sway with the rise and fall and bending of the flame to the wind, and that beckon the fancy out to grope in the mystery of night.
>
> Then imagination soars with the updrift of smoke and the climbing galaxy of fading sparks, to where the steadfast stars shine out of the unvisited realm that only imagination can explore.
>
> The camp-fire gives an expression to the human face that it bears in no other light, a vague intentness, an absorption in nothing tangible; and yet not a far-away look, for it is focused on the flame that now licks a fresh morsel of wood, now laps the empty air; or it is fixed on the shifting glow of embers, whose blushes flush or fade under their ashen veil. It is not the gaze of one who looks past everything at nothing, or at the stars or the mountains or the far-away sea-horizon; but it is centered on and revealed only by the camp-fire. You wonder what the gazer beholds—the past, the future, or something that is neither; and the uncertain answer you can only get by your own questioning of the flickering blaze.

What Robinson beheld as he gazed into the fire was mostly a nostalgic melancholy for the past, but there were times when the tongues of flame leapt for him into the prophetic realm of the future. His prowess as prophet derived from an ability shared by Thoreau and Whitman, and by nonliterary prophets as well—the capacity to See. This was no vague, mystical gift, but a hard-won prize, honed in the varying light of thousands of days spent attentively among men and women, and abroad on the Earth. Whether walking the cobblestone of South Street Seaport or along the banks of Sungahneetook, Robinson was always Seeing, always sketching, always noting the *particular*. In his field notebooks, when he hadn't his paints and canvas with him, he made valiant efforts at verbal descriptions of how certain trees or the entire landscape appeared in different types of light. "I must remember this in future winter pictures," he noted in one entry.

Thoreau's alertness to the visual came from his reading of John Ruskin and William Gilpin (the latter having expressed a tenet central to the work of both Thoreau and Robinson: "Language, like light, is a medium: and the true philosophic style, like light from a north window, exhibits objects clearly, and distinctly, without soliciting attention to itself"), while Robinson's sprang from his vocation as artist. The talent for close observation and expression came from his mother, Rachel Gilpin Robinson, who was an artist in New York City before she married Robinson's father.

When the artist's eye—Thoreau's or Robinson's—came to peer into the future, it had built up a deep reservoir of images intensely experienced, preparing it for prophecy. The "doomsaying" of both was the inevitable product of eyes that saw how diminished Nature was. Their gaze noted the number of wildfowl and songbirds fledged each spring, the fluctuations of wildflower populations, the health of streams and ponds. The numbers are there in their notebooks, and such are the numbers that impart to ecology the name "The Dismal Science." This part of Robinson's prophecy—the chronicling of diminution and the prediction of extirpation—was sounded early on in his writings. It rang out most dramatically (and sarcastically) in a piece entitled "A.D. 1950," which was published in the September 17, 1885 issue of *Forest and Stream*. The setting is unnamed, but it is unmistakably Lewis Creek as Robinson envisaged it sixty-five years ahead. This article was never included in any of Robinson's books, yet it is so "full of him" that it is reproduced here in its entirety.

> The usual September drouth was upon the earth. The grass was dun and slippery under foot; some of the small scattered trees wore a scant leafage of dull green, but more were burning in the smoky landscape with premature autumnal flames of yellow and red, casting blurred shadows under the redder rayless sun that burned its slow way through the brassy sky now endoming the dessicated fields and barren hills. Through a sedgy level a narrow, muddy stream wound sluggishly, too dull to reflect the blue spikes of pickerel weed that grew along its black shores, or the white blossoms of the sagittaria that drooped languidly over the spent arrows of their leaves. This slender waterway floated to-day a beautiful boat, with sides as thin as eggshell, but strong as steel, and buoyantly upholding its freight, a noble young gentleman of fourteen and his grey-bearded father, whom he had

kindly permitted to accompany him—indeed, had done him the unaccustomed honor of inviting him to go with him on this trial trip of his new canoe. The younger of the two guided the craft by almost imperceptible movements of his forefinger upon the tiller, while the elder propelled it with rapid strokes of the double-bladed paddle, and he, though laboring hard, seemed most of the two to enjoy the outing. He was revisiting scenes, that, though changed, were yet familiar, and recalled to his mind the past, concerning which he presently spoke.

"Right here," said he, as the canoe, under the skillful guidance of its master, lightly turned a bend of the narrow channel,"right here, when I was about your age, I killed my first duck with my first gun, which was only a double-barreled breechloader, a weapon somewhat out of date even then, but a great improvement on the percussion muzzleloaders our fathers had used, so we boys thought, for," said he with a sigh, "the males of the human race were called boys then till they were at least fifteen years old. The duck was of a kind now almost extinct, a dusky, or black duck, as we called them."

"*Anas obscura,* I think is the proper name," said the young gentleman.

"Ah, yes! *Anas, Anas,* what do you call 'em? And it was one of a large flock."

"Team! A team of duck would be better," suggested the younger.

"Very well; then one of a large team of five, and I brought him down with my second barrel."

"Ah, then you missed with your first? Why, I never miss."

"Those," the elder replied, "who never shoot, or never take any chances, seldom miss, though far be it from me to insinuate that this is why you never miss. However, we boys used to miss occasionally, while we were boys, but when we were older we scarcely ever did. I can remember very many hits, but few misses. ("How these old fellows do brag," thought the youth.) But this time, notwithstanding the miss, I was greatly elated, for it was my first duck, and I shall never to my dying day forget the place, just how it looked then, with the wavelets of the duck's plunge swaying the stems of the wild rice, and tossing the floating feathers, and the reflection of the trees that then lined the banks. And I remember too, with a twinge of shame, that I was not a bit conscience-stricken when the upbraiding look of the dying bird's eyes met mine. But that feeling comes later; boys are harder-hearted than men. A little further on, just around the next bend, one of those rare and most beautiful wildfowl, a wood duck, got up from the log where it sat sleeping in the sun."

"*Anas sponsa,* you should call it. Now that the game is gone we should treat the dear departed with proper respect, and I believe fewer ducks than a team are a plump, so you got up a plump of *Anas sponsa.*"

"Anything to oblige you, my son; but then, it was not plump, but thin as a shadow, as I found when I picked it up after bringing it down, as I was about to tell you, with my first barrel. Doubtless it had run the gauntlet of a hundred guns before it fell to mine. Then I was indeed jubilant, having got what many an older sportsman had failed to get, two ducks in one day—or perhaps I should say part of a team and a whole plump in one day—more than a sportsman of these days might hope to get in a season. Oh! but there were ducks in those days," the garrulous old man went on. "I have seen yokes; no, tandems; no, teams of a dozen and more right here on these marshes, where grew acres and acres of wild rice, and my father used to tell of seeing when he was a boy—for he, too, was once a boy—hundreds and hundreds in a—a—team. I suppose it might be said that these waters were then teeming with ducks. These waters, I say, for there were indeed waters here in those days all the year round. There were trees to shade them here and prevent their evaporation along these level reaches, and trees on the mountains that are now only bald wastes of rock, to shade the springs and keep the snow from melting all at once, to hold the moisture in the sop of deep moss and dole it out to the stream with a frugal hand. This waterway was twice as wide and deep as now, and full of good fish, pike-perch, bass, pickerel, with no end of perch and bullheads. Now only a straggler of the better kinds, some hardy survivor of his race, ever visits it, and the mudfish and gars have it all to themselves. Hundreds of muskrats built their huts of sedge and mud along the channel's edge in the fall, and found food in plenty at all seasons here, where now hardly a trace of one is ever seen. The great blue heron fanned his slow flight over the marshes that he is now a stranger to, and in the warm days of spring the guttural booming of the bittern sounded from every marshy cove. Now one might as well listen for the voice of the jungle fowl. Even the bullfrogs, since it became the universal fashion to eat them, have almost entirely disappeared, and the few that are left have grown so shy that one rarely hears the deep bass of a solo singer, much less the grand chorus of a hundred voices that of old used to roar along the sedgy levels and make them tremble with their loud melody—yes, and set the young leaves of the poplars aquiver on the distant hills."

"It must require a rather lively imagination to discover anything musical in the thin bass bellow of a bullfrog, I should think," the young man remarked.

"Well," his father replied, "it was pleasing to my uncultivated ear, as were the voices of the song birds then so abundant. In early spring were bluebirds—excuse me if I do not remember scientific names—the earliest minstrels of the year, 'shifting their light load of song from post to post along the cheerless fence,' as a poet of the last century says, and song sparrows singing songs of good cheer, with the jolly robins to help them keep our hearts alive with hope in the dismal days between winter and real spring. And purple finches atilt on the elm tops, and later the bobolinks, drunk with the wine of spring, singing as they staggered awing over the violets and dandelions of the May meadows. And flashing orioles that make one glad and sad with their song, and yellowbirds in summer, hanging like blossoms of gold on the thistles. They are almost all gone now. The stomachs of men and the bonnets of women have made way with them. It was the absurd and wicked fashion sixty years ago for women to wear stuffed birds on their hats, a fashion that raged so virulently that if a bird had handsome plumage or even shapely form, his sweetest songs and his prettiest ways could not save his life from the savage skin hunters who invaded all parts of the land, more cruel, rapacious and destructive than all beasts and birds of prey. And men, so called, ate robins, and even tickled their maws with the atomic carcasses of the beautiful snow buntings that used to come down in hordes from the north and give life to the white wastes of our wintry fields."

"Ah!" the young gentleman sighed, "they must have been very nice; much more delicate than crows, almost the only game birds one can get now."

"There were snipe on these marshes, a bird most excellent to eat," the old gentleman continued, "and woodcock, still more delicious, in the swamps and willow and alder copses that bordered the streams. The snipe were shot in spring and fall until, if there were any left, the marshes had become too dry for them to feed upon, and so they were exterminated or disappeared. The woodcock were shot in summer, and even in spring, and the swamps they haunted were cleared and drained to make meadows, the copses cut away to gain a few more rods forage, and so the woodcock were destroyed and banished. It was impossible to enforce the game laws, for they were cumbrous and full of loopholes that rogues, with the help of tricky lawyers, might escape through, and though liberal, they were held by the masses to be undemocratic, and after years of continual violation were abolished. In some places rich men leased large tracts of shooting grounds and protected the game on them. There were beside the snipe and woodcock, some ruffed grouse, a most noble bird now quite extinct, and wild quail, from which our domestic quail are descended. Also hares of two kinds in the eastern part of the continent, and red foxes. One or two individuals of this species are yet in existence, owned by the gentlemen of the Newport hunt, and are hunted by them every season. One of the foxes is so old and feeble that he has to be drawn before the pack in a low-wheeled cage, trailing behind it a bag filled with anise seed. The sport is said to be very grand and exciting, and nowhere else in America can one see this ancient and noble pastime pursued, with carefully bred hounds, blooded horses and well-equipped and gallant riders, and a real, live fox. The extinct possum and raccoon also afforded a great deal of sport, and were preserved in the leased tracts, and in them was found too the terrible skunk, the most formidable and dangerous wild beast of the continent. It secreted a potent fluid which it ejected by the gallon, paralyzing whoever inhaled its fumes, when the ferocious animal would spring upon and savagely bite its victim, who sooner or later expired in the agonies of hydrophobia."

"But these domains were eventually invaded by the march of improvement, the marshes drained, and all the woods cut down, for when the running bean became the fashion in Boston, every tree in the land half as big as your wrist was taken to furnish the immense bean fields with poles. Perhaps it was because the supply of poles was exhausted, and therefore the supply of the most brain-nourishing of beans fell short of the demand, that the ancient center of culture lost its distinguished position, and the continent became a great bicycle, as it were, Chicago the hub of its big wheel and Boston that of its little wheel, for you know, of course, as you know everything, that on the great plains of the West was conceived and put in practice the idea of training beans on sunflower stalks, by a wild man from England, which was not then a republic, for this was before the Battle of Dorking, you know."

"And here we are almost at the lake, or what was once the lake. Alas! it is only a great puddle now, or rather a series of puddles, and if the ancient explorer who named it could see it now, he would be sorry that he ever discovered it. But hark! There is a—a—I cannot recall the correct name, but what we used to call a helldiver, almost the only bird of its kind left. And there is a sportsman stalking him, and will soon open fire on him with his hopper-gun."

They watched a man clad in a suit of woven sedges and looking exactly like a great animate sheaf of the wild grass he was making his way through, till he arrived at the margin of the stream, where he set a tripod supporting a machine of curious construction. Into its hopper he poured a quantity of powder, shot and wads, and began turning the crank, whereupon the water for yards about the poor diver began to boil with the storm of leaden rain poured upon it by half a dozen revolving barrels, and for some minutes a succession of rapid reports filled the air. When these ceased and the foaming water became quiet, a cloud of feathers floated upon the surface, and were quickly brought to the sportsman by his retriever, till now unseen by the occupants of the canoe. Drawing near to the sportsman, they asked to see his game, and he with some pride showed his game bag half full of feathers.

"Thank you," said the elder of the canoeists, "but my son is something of a naturalist, and would be glad to see the bird."

"Oh," said the other, smiling,"there goes the bird swimming his best for the lake without a feather on his back. I did not wish to kill him, for he has furnished me a day's sport a year these five years, and if some bungler does not fall in with him, will give me as many more. I have picked him clean more than once, without injuring his bill, legs or eyes, or, I think, breaking his skin."

They went their way, marveling greatly at this shooter's skill. When they came to the lake their further progress was stopped by a sand or mudbar covered only by an inch or so of water. It was a dismal scene. The old rocky shores of the lake, once clothed with trees and washed by bright waves, stood now some furlongs inland, with a wide stretch of bare sand, cracked dried mud, and mossless, water-worn stones between them and the shrunken lake, whose turbid bosom no goodly fish broke into circling ripples, nor waterfowl swam upon. Not even a heron waded the black shallows, nor kingfisher clattered above them; not a sign of wild life was to be seen. The mountains to the westward were monstrous sterile piles of treeless rocks, savage and forbidding, not giving so much as a home to the eagle.

"Let us return," said the old man. "This is all so changed from what it was when I was a boy that I cannot bear to look upon it. The axe and fire—man's greed and carelessness and spirit of wanton destructiveness have spoiled it all. Let us go home, where we have at least an orchard and a well of clear water, and fields that are not entirely without greenness.

"O, why could they not have spared the trees upon these rocky shores, where they cumbered no tillable ground, and were so useful and so beautiful? The woods are gone, the waters are passing away and the hearts of men are grown as arid as the world they have spoiled. When I see how their lands are never withheld from laying waste the earth, from making sterile and forbidding all that was once so fruitful and fair, from exterminating nature, I cannot but be glad that in a few years my eyes will be shut forever from the sight of this 'abomination of desolation.'"

So, this is what the prophet of Sungahneetook saw just three generations ahead of him: "rayless sun," "brassy sky," "dessicated hills," "barren fields," fishless waters and birdless skies. He saw new technologies (the aluminum canoe and Gatling gun) in a dead and desolate landscape. Along with the heron and wood duck has gone the humor of man, his fictional son reducing animals to their binomials and finding no music in the chorus of bullfrogs. A.D. 1950 is now half a century past, a safe distance perhaps from which to evaluate Robinson's prophecy. What he saw has not been so much visited upon the Lewis Creek watershed, but if he had come back to see the "little wheel" (Boston) in 1950, Robinson would have undoubtedly believed his fiction to have come true. Some might interpret this futuristic extension of Rowland Robinson's lament as unreasonable nostalgia. Surely the lament that Robinson sounded in the 1880s had in the 1870s been voiced about the previous decade. There would always be some past "golden age" for every generation, and should not Robinson have learned a lesson from the Millerites, who awaited the end of the world on October 22, 1844, on hilltops and in roofless churches, but who woke up on October 23 to a world where eggs were still eighteen cents a dozen?

Even in the relatively pristine Lewis Creek landscape there are traces of Robinson's dark vision. Muskrats seem as fecund as ever, but the fate of the otter still hangs in precarious balance. Great blue herons labor out of the shallows in prehistoric flight back to their inland rookeries, but the tall trees that hold their enormous stick nests are increasingly encroached upon by new roads and houses. The rocky shore of Champlain does not stand "furlongs inland" from a shrunken lake, but each year fewer shorelines invite the wandering explorer, as they are developed for new homes. A flock of dusky ducks on Hawkins Bay is not a rare event, but it is an event that goes mostly unheralded, for though we have not become the humorless lout that rode in the canoe with his gray-bearded father in Robinson's story, we seem to *see* less. Aluminum canoes, Gatling guns, television, automobiles, and our own uncertainty of our place cloud our vision. Our forefinger is on the tiller, yet we still have not fathomed where we are headed.

From time to time when Robinson was out on his rambles along Lewis Creek and Little Otter, he would be absolutely paralyzed by the sheer beauty of a scene. On October 28, 1881, on a crisp day when the air was like cool spring water and the swamp maples burned like fire on the hills, Robinson made this notation as he walked along Lewis Creek: "What *is* the feeling one has in seeing a beautiful thing, landscape, single object, or figure? A pleasurable *ache* to absorb it, to in some way describe it in words or pictures, an *ache* of inadequate expression?" That ache, the painful resonance between the outer world and his very bones, was a symptom of his capacity for seeing, and feeling into, the modest Champlain Valley landscape that evoked mostly utilitarian sentiments from his fellow Ferrisburg farmers. Given his sensitivity to the fabric of life that entwined the Creek, it is understandable that Robinson transcended the mores of his times. It also explains how in his last dozen years of painting prose portraits of his "three rivers," he no longer needed to physically see them. In 1887, Robinson's eyesight began to fail. As a child, someone had poked him in the eye while combing his hair, and that eye had remained weak. In his fifties, blindness threatened. Friends and relatives offered advice, all of which Robinson accepted gracefully. In 1888 he received from his friend George Washington Sears—known to readers of *Forest and Stream* as "Nessmuk"—a letter prescribing a "hemlock and scraped potato embrocation" to restore sight. Other friends suggested surgery, and Robinson traveled to New York for a number of operations. All of these proved unsuccessful, and in 1893 he became totally blind.

As so frequently happens, another sense compensated for the visual handicap. The blind artist's hearing became more acute than ever, and his creative vision became lodged as much in his ears as it had been in his eyes. Now he sought solace in music, playing the alto horn he had used in his youth, and singing more fervently than ever. He still loved to smoke a pipe, and could with ease strike his own match, for he used a ground slate gouge for a striker. He had found the ancient artifact weathering out of a bank near the mouth of Lewis Creek, and its smooth channel made a convenient and easy-to-find surface with which to light his pipe. His writing became his constant comfort, and with his new aural acuity he refined his use of dialect, writing on a grooved writing-board developed by the Perkins Institution for the Blind. He would follow along the groove with his left forefinger behind his writing hand. After each page of manuscript, his wife or daughters would read it aloud to him, making corrections if necessary. Cut off from the visible world, he lived in his memory and in his imagination. By his art he made these visible, and audible, to others.

Rowland Robinson could record Abenaki place names for the same reason that he could write dialect; he had a marvelous ear. That gift was most easily understood by those around him, the family and friends who were treated to his fund of anecdotes and stories. Sitting by the hearth at Rokeby, telling a story about frontier life in the Champlain Valley, he would make the wild howl of the wolf, the sound of the wind in the tall pines, and reproduce faithfully the voices of Indian, Scottish settler, and French-Canadian alike. He also sang beautifully, in a sweet, though untrained, voice, and his children heard songs from him that they never heard anyone else sing. People beyond Ferrisburg knew of his gift from his books, which, though difficult to read because of the extensive passages of dialect, were read aloud in thousands of homes across the country. Transplanted Yankees from Florida to California grew homesick reading Robinson's Danvis tales.

His use of dialect may have restricted Robinson's popularity both in his time and in ours, yet it is the principal strength of his writing. His descriptive prose is, like much of the writing of its time, uncomfortably overwritten, but in dialogue he finds a voice that rings true and clear. Poet Hayden Carruth, comparing Robinson to the most famous American dialect writer—Mark Twain—found Robinson's work to be:

> far from the paler construction . . . he never condescends to his dialects, as he never condescends to his characters. His attitude is serious and affectionate. The dialects he records are real languages, still close in many respects to the Tudor speech of the first New England settlers. Robinson appears to have sensed this, though he was not a linguist—he could not have been in his lifetime—or even a philologist. But he was a writer, an artist working with language, and he had an ear for sound and cadence. He knew the speech of his region, he knew what it meant, he knew the people who used

it. He recognized the bedrock integrity and human value of that speech, and this at a time when the dominant gentility of America insisted that such speech could be nothing but vulgar.

Walking was still a joy to Robinson, and after he became blind he went with a strong hickory staff in one hand, his left upon the shoulder of one of his children or a friend. His eyes remained clear blue and full of expression, and when he spoke he almost always looked into the eyes of the person with whom he was speaking. When he moved about the house, he did so with such an easy, straightforward manner that the casual observer did not discern his blindness.

Still, Robinson felt a great loss. In October of 1893, just after his sight had wholly gone, he wrote to a friend:

> Your letter came just as I was unfolding my bat's wings for a first flight in the dark. I 'lit' twenty-four miles off, and after a week's very comfortable visit sailed home again yesterday. I would rejoice to get a letter from . . . you . . . but I shall not have the chance of rejoicing. "I'm Dead, I'm dead," as my schoolfellow, Charley R., shouted when he got a licking; and I know something how it seems to a man to hear the world going on around him, and he lying quiet under the grass.

Robinson found it difficult to tell his many friends of his loss. When he did so he tried to stay his melancholy, but the response was inevitable. Occasionally his correspondent would be more philosophical than emotional. Willis Royal Peake, a friend from Bristol who was also going blind, sent Robinson an article about x-rays being used to cure blindness, and encouraged him to be hopeful. Those who truly understood him, however, could not shake the feeling of tragedy. Charles Faxon, with whom he had botanized all over the Lewis and Little Otter Creek watersheds, wrote:

> Never have I felt such mingled pleasure and pain as your letter has given me . . . your written words were the first intimation I had of your terrible affliction. I began your letter with all the zest that the remembrance of some of the most delightful hours of my life awakened, only to receive a cruel shock at the close which leaves me scarcely able to send a few poor words of sympathy. I am aware how weak such words seem but they are all my heart can find to express its sympathy. I read between the lines that you bear your cross with heroism. Thank God you have wife and children to love and comfort you and that you are well furnished with food for the *mind's* eye. Happy they who in your care can draw on a memory well-stored from the treasures of literature and art.
>
> I told them of our tramps together in dear old Vermont. More happy days!

Robinson did have more happy days. In the last seven years of his life he wrote nine books, and continued his correspondence. He still walked, though he was confined to his own farm, and his old hunting and fishing buddies often took him to the landing on the East Slang, for canoe trips up Little Otter to Hawkins Bay and Lewis Creek. His field notebooks were a little larger now, so that they might accommodate the corrugated writing guide between their pages. The penciled script stood straight up on the page and sometimes leaned illegibly back on itself, but the observations were as clear as ever—remarks on rattlesnakes that used to frequent the hills east of Monkton Pond, where the best places were to fish for eels, and notes on how the Abenaki trapped beaver.

One of these outings was immortalized in the February 1895 issue of the *Atlantic Monthly* as "A Voyage in the Dark":

> A few days ago, a friend who is kind and patient enough to encumber himself with the care of a blind man and a boy took me and my twelve-year-old a-fishing. It was with a fresh realization of my deprivation that I passed along the watery way once as familiar as the dooryard path, but now shrouded for me in a gloom more impenetrable than the blackness of the darkest night. I could only guess at the bends and reaches as the south wind blew on one cheek or the other, or on my back, only knowing where the channel draws near the shore upon which the Indians encamped in the old days by the flutter of leaves overbearing the rustle of rushes. By the chuckle of rushes under the bow, I guessed when we were in mid-channel; by the entangled splash of an oar, when we approached the reedy border.

On this voyage the blind man recognized many of his acquaintances—the raucous kingfisher, a roar of redwings out of the wild rice, a croaking bittern, and wood ducks whistling away over Hawkins Bay. He knew by the smell of cedars when they approached Gardiner's Island, and felt his way familiarly from the Slab Hole sandbar up to the old camping place on the ledge above. His son delighted to see the spot, with its tiny caves bedecked by Canada yew and mountain fringe: "Through his undimmed eyes I had glimpses of those happy shores whereon the sun always shines and no cloud arises beyond. What a little way behind they seem in the voyage that has grown wearisome, and yet we can never revisit them for a day nor for an hour, and it is like a dream that we ever dwelt there!"

As wild apples, "naturalized" lovers of the untamed landscape like Thoreau and Robinson were struck through with self-cultivation. Robinson's blindness, like Thoreau's own personal losses, only increased the depth to which he tilled the soil of his inner self. The scruffy orchard near the railroad trestle over Lewis Creek tells us that we too are wild apples, and that although most of us may escape the challenge of the loss of our physical eyesight, each of us daily risks losing our larger vision. The land, like personal tragedy, calls us home to ourselves, and in going there, we stay the course between a comforting past and a frightening future. Rowland Robinson's prophecy of a depauperate fauna at the Lewis Creek marsh has thankfully passed unrealized, but his subtle warnings of an inner poverty—personified in the literalist language of the young boy who sees nature a bit too scientifically—remain poignant. Like the wild apple, we moderns who have survived A.D. 1950 and have just moved into a new millennium must express our finest qualities, offering our fruit to the future as those wild apples before us have done.

CHAPTER 5

. . . and Weeds

It is quite like actual farm experience, that among so much good grain as we have here there should spring up something weedy.
—Cyrus Pringle

HE RAILROAD, LIKE the river, is still an avenue for prophecy, for glimpses of things to come. The steel rail's ballast banks harbor a small, mostly unseen advent in the weeds that grow there. Scramble up the man-made bank, held in place by a weedy riot of plants, to the tracks, and walk with them north or south. A distinct plant community lies at your feet, paralleling the paired rails, diverging slightly in its composition from time to time along the railroad's length, but characterized by its collection of tough aliens.

Not that all of the railroad's weeds are nonnatives. In early July a walk along the railroad would show a number of plants common to the banks of Lewis Creek: yellow wood-sorrel (*Oxalis europaea*—the specific name is a misnomer), whose leaves and seedpods make tasty munching on a hot July day; milk-purslane *(Euphorbia supina),* a little spurge that lays in flat mats on the stones; and daisy-fleabane *(Erigeron annuus).* A congenitor of the last, the robin-plantain *(Erigeron pulchellus)* sometimes makes its way up from the purple-flowered colonies on the river banks below to dot the railroad with color in May. These riverbank natives, which evolved to contend with the spring flooding of their riparian habitat, do equally well on the exposed gravel of the railroad right-of-way.

But if a native plant can be called a "weed," what is a weed? Cyrus Pringle the farmer recognized the old definition of a weed—"a plant out of place"—but Pringle the botanist acknowledged no weeds, since to him "every plant, whether it possesses any economic value or not, is, in its peculiar and marvelous structure, an object of interest." Pringle knew the weeds of the Champlain Valley as well as anyone, and much of that knowledge came from walking the railroad between Charlotte and Ferrisburg stations. While he collected the weeds of the Lewis Creek and La Platte watersheds, his friend Ezra Brainerd was hunting weeds along the track in Middlebury. They both knew this linear, man-made habitat as one of the best places to hunt weeds, for not only was the habitat conducive to their establishment and growth, but trains every day brought plant aliens that might drop off and establish themselves in the ballast. A passing train might carry thousands of unpaying plant passengers on livestock, packing material, and even the clothing of those who had paid their fare. They came from all sorts of exotic places, and got off usually at the loading yards, making their way along the tracks from there. In the old floras such plants were called "adventives"—introduced but not fully naturalized, not unlike the Irish men who laid the tracks. Only time could tell whether the immigrants would stay, as many of those found by Pringle and Brainerd have. Despite the decline of the railroads in Vermont, it still brings newcomers; just thirty years ago, a Massachusetts botanist discovered three plants new to Vermont in one afternoon in the railroad yard in Burlington. One, *Elsholtzia ciliata,* was a mint native to Asia, while the other two were grasses from the southern United States—*Aristida intermedia* and *A. oligantha.*

One hundred years ago, some of the plants still seen along the railroad were first making their appearance in the watershed, and in the state. From the beaches around Boston came orach or spearscale, *Atriplex patula* var. *patula.* From Europe, via Boston, came a relative of *Atriplex,* the oak-leaved goosefoot, *Chenopodium glaucum,* and two mustards, *Lepidium campestre* (field peppergrass), and *Sisymbrium altissimum* (tumble mustard). From the midwest came the symbol of peregrination, the tumbleweed *(Amaranthus graecizans).* Adventives rode the rails south from Montreal also, such as biennial wormwood, *Artemisia biennis.*

Some of the plants that "rode the rails" were wonderful surprises to Pringle, Brainerd, and their fellow botanists. For Pringle, the prickly poppy, *Argemone mexicana,* must have been such a find. In its variegated armored form, leaves turned entirely into stiff, sharp spines, it held another landscape. The desert was there in the basal rosette of that poppy. If he had listened carefully as he bent to admire the hypnotic radial array of unfolding leaf-spines, this adventive might have told him of remote limestone canyons, of dry desert washes where only the toughest plants survived. He probably never dreamed that the railroad would take him to the home of this desert *yerba,* but it did. On April 2, 1885, Pringle and a young farmhand from Charlotte, George Welcome, hopped off a train in Bachimba Cañon, about thirty miles south of Ciudad Chihuahua, Mexico. Just three months before, Pringle had written to Harvard's Asa Gray: "I am jubilant this morning. With this pass by which I may travel at will along the base of the Sierra Madre, and an unusual depth of snow in these Mountains to ensure a good growth of vegetation the coming season, I feel that the success of my next journey is almost assured."

Though the Mexican Central Railway, which ran from the border city of Juarez south to Chihuahua, had its offices in Boston, and though Gray worked with his most influential Harvard contacts to obtain a pass, his efforts were stymied. "Everyone wants to go to Mexico, or talks that way," said the Railway president in response to Gray's request. Gray had fought and won much larger battles, most notably as Darwin's champion against Louis Agassiz's special creation view of evolution, and this loss stung him. After a chagrined and humiliated Gray wrote Pringle in early 1884 of his failure to secure passes, he regrouped and tried again. By January of 1885, he had them, one for Pringle, and one for an assistant. In return, Pringle was to make a report to the railroad on the "vegetable products and especially the timbers of Chihuahua." Reliable sources of lumber for construction were of utmost concern to the railroad entrepreneurs, especially in an arid region like the Chihuahuan Desert.

The railroad pass was a prize hard won by Gray and his influential Harvard associates, and gave Pringle a route to botanically unexplored lands south of the American border, lands that Gray desperately desired to see through the eyes of his most indefatigable field collector. Though on this trip Pringle never made it to the Sierra Madre, his explorations in the hills near Chihuahua City brought ample botanical rewards, rich enough that the retiring Quaker from Vermont returned to Mexico each year until his death in 1911. His twenty-six years of botanical wanderlust radically altered scientific knowledge of the North American flora—10 percent of the more than 12,000 species of Mexican plants that he collected were new to science.

Pringle and Gray's enthusiasm for a Chihuahuan collecting expedition began in 1884, when Gray had Pringle working in the Santa Rita Mountains of Arizona, from whose heights he "first beheld Mexican territory, the rugged heights of northern Sonora receding crest upon crest in paler and paler blue beneath the staring southern sky—a land of mystery and fear." The work in Arizona was perfect training for the Vermont collector—the 683 species from his 1884 Arizona trip were decidedly Mexican in character. Gray, whose *Synoptical Flora of North America* did not include this region, was delighted with Pringle's efforts there. He described most of Pringle's 1884 discoveries, and named one genus after his faithful collector—*Pringle-ophytum*—"for his very arduous and hazardous excursions made . . . where no botanist had hitherto penetrated."

Perhaps the most hazardous element of travel in that region was the presence of warring Apache Indian bands, led by Geronimo. Edward Palmer, a fellow collector active in the Southwest, expressed surprise when he heard that Pringle was collecting in Arizona, since, when he had been there recently, he had needed a military escort to go anywhere. Pringle's success in Arizona only whetted Gray's appetite for new lands and new botanical material to describe. In letters to Pringle, he stated emphatically, "I want you in *new ground*!" The railroad carried Pringle to the new ground, and transported him from habitat to habitat once he was there. From his base in Chihuahua City he rode north to the Samalayuca Dunes, east via the Santa Eulalia Mining Company's narrow-gauge railway into the rugged limestone canyons of the Sierra Santa Eulalia, and south to Bachimba Cañon. The Mexican Central Railway there follows the Rio Bachimba through a set of sandstone hills that rise above the desert floor. Tall cottonwoods dot the narrow floodplain along the river, which in the dry season of Pringle's arrival, ran as a small brook. The aspen-green of the cottonwood leaves and the ribbon of water must have looked inviting to a Vermonter abroad in the Chihuahuan Desert for the first time. With letters of introduction from the governor of the State of Chihuahua and the general manager of the Mexican Central Railway (secured with letters from Gray and Spencer Baird, Secretary of the Smithsonian Institution), Pringle prevailed upon the conductor to halt the train near the trestle over the Rio Bachimba.

In Bachimba Cañon Pringle collected his first new species of his Mexican collecting expeditions, *Anisicanthus insignis* Gray. He found a number of ferns he was familiar with from the previous year's exploration in Arizona, a pretty milkweed, *Asclepias nummularis,* and an Acanth, *Nolina texana,* collected in 1884 as number 1 of Pringle's numerical list that would

Cyrus Guernsey Pringle and his field assistant Filemon Lozano collecting in Mexico. Courtesy Pringle Herbarium, University of Vermont.

eventually number over 12,000 species. In a dry wash between the rail line and the river, he also collected number 257, *Argemone mexicana,* the prickly poppy. Called *cardo mano* by the Mexicans, it was one of the most familiar of desert plants, gathered frequently for medicinal purposes, and sometimes even cultivated. Its fate was as closely connected to the railroad as Pringle's, with the steel rails being the avenue for both of them to new habitats, and new fortunes.

If you stray off the railroad right-of-way and across the fields that still crowd the Creek north and south here near the trestle, you will find the same weeds vexing the farmers as they did in Pringle's day. In pastures there are shepherd's purse, *Capsella bursa-pastoris,* which was rated by an international group of botanists the 26th worst weed in the world; common St. Johnswort, *Hypericum perforatum;* the fleabane met on the railroad, *Erigeron annuus;* dandelion, *Taraxacum officinale;* and a number of umbels—Queen Anne's lace, *Dauca carota;* caraway, *Carum carvi;* and parsnip, *Pastinaca sativa*. A variety of grasses, mostly European, still thrive in these same pastures. One in particular, *Triticum repens,* was and is especially infamous with farmers. Its common names suggest how widespread and detested the plant is by farmers—couch, quack, quick, bunch, joint, snake, witch, and devil's grass. Pringle called it "the most troublesome and obstinate pest I encounter in farming," and on one occasion, when he found a patch of it in his pear orchard, he and a hired man spent three days digging with a spading fork and potato hook about twenty bushels of roots. The following spring he harvested two bushels, and a year later but half a bushel. He attacked devil's grass early and often, believing that the old adage, "give the devil enough rope and he'll hang himself," did not apply at all to *Triticum repens,* which Pringle declared not perennial, but "*eternal.*" Ferrisburg farmers now wage their war on the tough grass with herbicides instead of hoes, but it is equally vexacious to them.

In Pringle's day some of these fields grew grain or flax, and both crops had their own associated weeds. In wheat fields could be found charlock or wild mustard *(Brassica kaber),* cockle *(Agrostemma githago),* and chess *(Bromus secalinus).* The flax field invariably contained false flax *(Camelina sativa).* While these crops have declined, and with them their weeds, other cultivated habitats persist from a century ago. Wet meadows on the farm still host a suite of buttercups (*Ranunculus* spp.), meadowsweet *(Spiraea alba),* steeplebush *(Spiraea tomentosa),* meadow rose *(Rosa blanda),* and Joe-pye-weed *(Eupatorium purpureum),* all native plants but "out of place" to the farmer bent on cultivating every inch of his land. Dryer spots usually sported common cinquefoil *(Potentilla canadensis)* and sweetbriar or eglantine *(Rosa eglantera).* On the clay of many Champlain Valley farms, a succession of weeds could be seen. The commonest plant on the clay—timothy *(Phleum pratense)*—was itself out of place, being a native of Europe, but desirably so to the farmer. In many places, Pringle's peers found that the nutritious grass was being supplanted by ox-eye daisy *(Chrysanthemum leucanthemum* var. *pinnatifidum).* Some farmers thought that the daisy "ran out" the grass, but Pringle recognized the failure of timothy to be due more to severe droughts (of which there were a series in the 1860s to the early 1870s) and frosts, and to the partly exhausted condition of the land. His experiments showed him that phosphorus was the critical nutrient for timothy, and that the valley clays were particularly deficient in it. He believed the daisy to be no great nuisance, and even suggested that if cut when in blossom and properly cured, it made sweet and nutritious hay.

It proved a reliable backup in the drought years when the hay crop was lightest, and pragmatic Pringle the farmer admired how "over the fields of the Champlain Valley it spreads in June its white banners." Even the daisy sometimes could no longer find sustenance in some fields, and then it was succeeded by *Antennaria plantaginifolia,* the plantain-leaved everlasting, or, as the farmer's of Pringle's day called it—"mouse-ear." "It is the last vegetation with which these lands can robe themselves," said Pringle, "and the ghastly hue of the plant is appropriately suggestive of the condition of the soil beneath it."

Pringle must have haunted door-yards as often as he tramped through fields and along the railroad, for he found dozens of adventives newly escaped from herb and ornamental gardens. Some of them were common garden escapes—teasel *(Dipsacus sylvestris)*, tiger lily *(Lilium tigrinum)*, day lily *(Hemerocallis fulva)*, tansy *(Tanacetum vulgare)*, live-forever *(Sedum telephium)*—while others were rarer—garden burnet *(Sanguisorba minor)*, henbit *(Lamium amplexicaule)*, and spotted dead nettle *(Lamium maculatum)*. Pringle found in his own door yard a plant native to Eurasia, summer-cypress *(Kochia scoparia)*, which was the only place he or any other botanist ever found it growing in Vermont.

The ultimate fate of those aggressive foreigners, whether they hung around kitchen gardens or doorsteps or railroad yards, rests with humans. Like domesticated crops, their increase in this watershed so far from their native lands can come only with human nurturing. Like varieties of apple trees, they have and will continue to fall in and out of favor with the people of their adopted home. In 1872, Cyrus Pringle said about chicory, *Cichorium intybus,* that "its recent and rapid advance upon us should excite alarm. It is believed to impart a bad taste to the milk of cows feeding on it." Today the showy blue flowers of chicory are to be seen on every roadside in the Champlain Valley, and it is even one of the main species in the wildflower mix spread by road crews along stretches of the interstate highway.

In the 1870s, when Pringle was lecturing farmers throughout the state about pernicious weeds, he made no mention of the orange hawkweed, *Hieracium aurantiacum* (also known as devil's paint-brush). He had seen it before; in 1873 he came upon a cluster of it by the roadside near Shelburne Falls. The showy native of alpine regions in Europe was introduced into the eastern United States as early as 1818, but it had spread very slowly until around the time when Pringle collected it. A principal reason for its increase then was that an agricultural journal offered it as a "premium." By

1891, when the Vermont Agricultural Experiment Station made a partial canvass of the state to identify Vermont's worst weeds, hawkweed ranked fourth, just behind quack grass, ox-eye daisy, and charlock. The Experiment Station's rhetoric made the pretty orange composite sound like the plague. "It now stands unequalled as a dangerous weed in some sections of Vermont as shown by *its rapid spread, utter uselessness and thorough monopoly of the soil.*" Handbills were printed up and distributed throughout the state, advising farmers and homeowners to apply salt to the dreaded alien. Its fecundity was something to be feared, said the literature; not only did the plants spread vegetatively (thirty new plants originating from a single old plant), but each plant blossomed almost continuously from June to November, producing over a thousand winged seeds.

In the early 1900s, Pringle and Robinson's farmer neighbors fretted over the loss of precious nutrients to unpalatable invasives. The state Department of Agriculture put out bulletins to stop their escape and proliferation. Amateur botanists and village beautifiers throughout Vermont went on the warpath against weeds, organizing special exhibits at county fairs on the threats posed by weeds. That all this anxiety over weeds flourished at the same time as the same group of New Englanders—educated, "progressive" men and women—grew enthusiastic for immigration reform laws was no mere coincidence. Turn-of-the-century Yankees were being tumbled about by change, and their efforts to freeze, at an idealized moment in time, both their physical and social landscapes signified their fervent desire to stay those changes. This was the same historical moment that saw Wallace Nutting and other New England antiquarians mythologize a pure and homogeneous Yankee past. They created the mythic image of Vermont villages as universally of white clapboard houses on tidy greens, with white Protestant folk inside the well-kept houses.

After all the advice on how to eradicate the orange hawkweed, the Experiment Station said in 1897, "The hawkweed is in Vermont to stay." Here was true prophecy. Walk from the trestle over the Creek to the Pringle homestead, and you'll see it continuously. The few farmers left in Charlotte and Ferrisburg still think of the plant as a nuisance, but they are outnumbered by the newcomers with neatly mowed lawns who don't mind the mid-summer splash of color, and who steer around them, unmindful of their "utter uselessness." The hawkweed and other fellow invaders of the Lewis Creek watershed have lately begun to receive the sort of acceptance that Rowland Robinson bestowed on runaway slaves and emigrant Quebecois. However "alien," these floral invaders now call forth not fear but friendship. Or at least the tentative first steps toward friendship. While ten years ago the nonnative purple loosestrife was the scourge of wetlands lovers, today it seems to call forth a "let's see if we can get along" attitude from some of those who would conserve and restore wetlands.

"Conserve" was the byword of Robinson and Pringle's generation; "restore" was as yet a rarely practiced relationship with the land. At the close of the twentieth century, we are realizing the philosophical complexity, and ambiguity, of ecological restoration, at the same moment as conservation biologists are refining the necessary tools and techniques to accomplish restoration. Weeds like purple loosestrife, so recently marked for removal, suddenly have symbolic significance within the new dynamic view of ecosystem processes. One persistent weed that came into the Lewis Creek watershed after Cyrus Pringle's day is Japanese honeysuckle. The naive dayhiker may pass it unnoticed, but to gardeners throughout the Champlain Valley, it is a menace, its prolific seed production and distribution and vegetative spread allowing it to quickly dominate native plants. A decade ago it would have merited nothing but scorn from ecologists and native plants enthusiasts. Yet today, at the mouth of the La Platte River, the next watershed north of Lewis Creek, Nature Conservancy land managers are looking at Japanese honeysuckle as a rightful immigrant. Disabused of earlier notions that the Conservancy could simply purchase a piece of land and restrict human use as a way of keeping the land pristine, land managers now see the Japanese honeysuckle as an integral part of a never-ending process of historical change.

A century has given us time to see that Cyrus Pringle's pragmatic observation "that among so much good grain as we have here there should spring up something weedy" should not stand as a statement of resignation, but of celebration. Weeds and wild apples speak to us of nature's persistence, tenacity, and immense restorative powers. They tell us that "this too shall pass" and that whatever comes next, though it may not have been in our plans, is bound to hold surprise.

CHAPTER 6

Star in a Stoneboat

But a botanist's experience is full of coincidences. If you think much about some flower which you never saw, you will be pretty sure to find it some day . . . growing near by you. In the long run, we find what we expect. We shall be fortunate then, if we expect great things.
—Henry Thoreau, in a letter to Mary Brown of Bennington, Vermont, May 1859

Never tell me that not one star of all
That slip from heaven at night and softly fall
Has been picked up with stones to build a wall
—Robert Frost, "A Star in a Stoneboat"

HE 1871 BEERS *Atlas* map of Ferrisburg is a good companion when reading Rowland Robinson's journals, or one of his books. It shows the location of Rokeby, the old Academy ruins, all of Robinson's neighbors, the North Ferrisburg railroad depot, the schoolhouses, and the Quaker cemetery. It even locates a mineral spring behind the Robinson homestead, a "black marble quarry" near the shore of Hawkin's Bay, and a lime quarry west of Fuller Mountain. Then there is a more puzzling notation just east of the Louis Fuller farm—"Meteoric Rock."

It has been over a hundred years since Louis Fuller or one of his neighbors mentioned the fallen star to the Beers survey man as he walked his wheeled measuring device over the dirt road past the Fuller farm. He must have even stopped to see the strange thing, since the *Atlas* notes its location

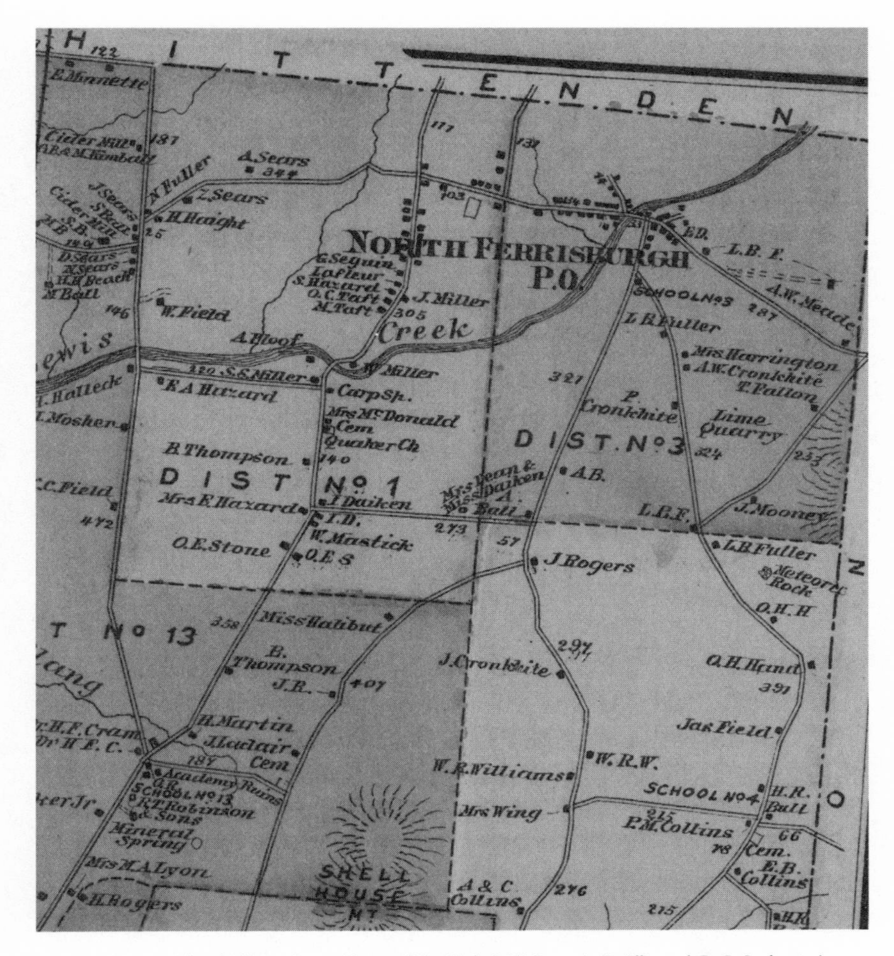

From F. W. Beers, *Atlas of Addison County, Vermont* (New York: F. W. Beers, A. D. Ellis, and G. G. Soule, 1871).

as faithfully as it could. To such earth-tethered men the rock must have been marvelous. They had seen the Perseid shower each August, and they well-remembered the Leonid shower which had lit up the sky in November of 1867. That one of those brilliant lights had landed on his farm was a never-ending source of wonder to Louis Fuller.

Stop by the Fuller homestead today, and the present occupants can tell you right where the "meteoric rock" lies: "Follow the fence line east of the house to a little patch of hardhack and wild cherry. It sits right in the middle of those trees." It's larger than you imagined, almost ten feet high and more across, and yet there it sits, no boardwalk around it, no plaque commemorating it. Its surface is pitted and pockmarked with all shapes and sizes of cavities, and thin ridges stand up in some places on it. It is buff-colored, but in some spots it's black or gray or red, or green where moss covers it. It is a strange-looking rock to find in a Ferrisburg fencerow, but it is no meteorite. The fallen star is really a glacially strewn boulder of Dunham dolomite. The myriad surface holes, so archetypally meteoritic, are not the escape vents of stellar gases, but solution cavities, where earthly water has had its way on soft calcium carbonate. The black, iron-like look of the stone comes not from dense metals forged in outer space, but from a few thousand years of weathering and a few species of lichen. The curious ridges contain not the stuff of star creation, but of earthly decay. They are layers of indurated sand that were laid down on some shallow Cambrian shore amongst the magnesian limestone.

Older residents of Ferrisburg remember hearing stories about the meteorite. Whether it really was or wasn't one they couldn't say, for no expert ever came to tell them. Had word of the Fuller farm "meteorite" reached a scientist in this century, it seems it would have been investigated, for the number of known meteorites from New England was and still is extremely small; in fact, none are known from Vermont or New Hampshire. Where did they all go? Were they dropped anonymously into stone boats, like the subject of Frost's poem?

> It went for building-stone, and I as though
> Commanded in a dream, forever go
> To right the wrong that this should have been so.

Though a false fallen star lives in memory and on a map, real ones are forgotten. In March of 1877, a man from Waterville, Vermont, wrote to the editor of the *Burlington Free Press:*

> At 5:50, on the morning of March 20, while at work in the front yard, I noticed a peculiar bright light, and looking up, saw what resembled a meteor or an acrolite, start from near the zenith, and passing north of northeast to near the horizon, vanished, leaving at first a stream of light, then a blue line of cloud closely resembling the smoke of a rifle, which faded entirely in about 30 minutes, passing away toward the south. A stiff north wind was blowing, and the thermometer indicated minus four degrees. After a lapse of about 3 minutes, a rumble as of heavy artillery, accompanied by a slight jar of the earth, was heard and felt, which lasted about 60 seconds.

Other people in Waterville saw the flash of light produced by the meteor as it entered the earth's atmosphere, and heard what they believed to have been thunder. In Newport, Vermont, some people thought there had been an earthquake. Two weeks later, a party hiking up Jay Peak came upon what they at first took to be an explosion:

> The snow had the appearance of blue vitriol, with a sulphurous smell, and a soft substance like soap or sugar covering it for a space of 4 or 5 rods. In the center there seemed to be an excavation in the snow and earth. On examination they found imbedded in the earth about 4' below the surface, something like a conglomerate rock, having the appearance of iron ore, lava, and soapstone, mixed with a substance similar to blue glass. The stone is supposed to weigh about 2 tons, and will probably be taken to Jay Peak, for exhibition the coming season. The great wonder is where did all this 'calamity' come from?

Talk to people in the Jay Peak area today and the meteorite story draws a blank, yet in Ferrisburg, legend still survives about a dolomite boulder in a hardhack hedgerow being a fallen star. Geologists today agree that there are no Vermont meteorites, and some speculate that because of the region's long history of settlement, they have all ended up in smithy's forges or stone walls.

Robert Frost was living in New Hampshire when he wrote his poem, "Star in a Stoneboat," which tells of a star that "softly" slipped from heaven. Like many of Frost's poems, it pits poet against pragmatist. The farmer rudely pokes the fallen star with an iron bar, and after a brief ride in a stone boat, into the wall it goes. Frost recognizes that it is as suitable a resting place as any:

> Yet ask where else it could have gone as well,
> I do not know—I cannot stop to tell:
> He might have left it lying where it fell.

If he had left it lying where it fell, there would have been no poetry in it. Frost's real meteorite, like Louis Fuller's imagined one, was a marvelous thing:

> Though not, I say, a star of death and sin,
> It yet has poles, and only needs a spin
> To show its worldly nature and begin
>
> To chafe and shuffle in my calloused palm
> And run off in strange tangents with my arm
> As fish do with the line in first alarm.

Had Frost been the surveyor who was taken out to see Louis Fuller's "meteoric rock," his poem might have told the same tale. Though too big to pry onto any stone boat, the Fuller meteorite took its place between two fields. Though no fallen star, it encompassed worlds as surely as Frost's farmer's true meteorite did.

> Such as it is, it promises the prize
> Of the one world complete in any size
> That I am like to compass, fool or wise.

A mile past the Fuller farm the Beers surveyor walked by a landmark as momentous as the meteoric rock, but paid it not the slightest bit of heed. On the Heman Bull farm, just south and east of where he noted a small tributary of Lewis Creek passing under the road, a stand of Vermont's rarest tree—*Pinus banksiana*—stood. But it may as well have been a real meteorite, for no one, not even Heman Bull, knew of its existence. Like the thousand fallen stars mute in sugar woods and cedar swamps, the grove lay hidden, though fully in view. Red squirrels knew of it, for they tore open the twisted cones year after year. Deer knew of it, yarding up under the gray pines in winter when the snows were especially deep. Grouse knew of it, browsing the buds on their foraging rounds. No farmer or surveyor or botanist knew they existed, so they did not exist. There were no "jack pine" or "gray pine" in Vermont.

Until September 27, 1876. On that day Rowland Robinson recorded in his journal: "This day I found on a ridge north of 'Ruey Hill,' a little colony of *Gray or Lerut Pine,* the first and only I have ever saw." Twenty years later he set down a more detailed record of his discovery in a letter to his friend Charles Faxon:

> It was not less than twenty years ago that I first came upon the group of gray pines, one day in November. I was fox-hunting and fox and hound had gone over onto the Monkton hills farther than I cared to tramp so I loafed around in sunny places listening and waiting for their return to the Chronk hill. This is the rocky hill east of the south end of Shellhouse Mountain and was named for a disreputable old Quaker who lived on the west side of it. A committee of Friends was appointed to "deal" with him for getting drunk and after they had "labored" with him awhile he told them that he always 'sot gret store by that passage o' Scripter that says "Let every tub stan' on its own bottom.'" Larrence, as he was always called though his name was Lawrence, was incorrigible and Friends felt obliged to disown him. The name of this family is an odd instance of change. It used to be spelled Cronkhite and

so pronounced by them and all others who had a proper regard for correct speech but now they are Cronk to themselves and all the rest of the world and the old name is rarely spoken or written.

But all this has nothing to do with the gray pine unless Larrence got his bad luck by going near the group which grows on the next ridge east of his hill.

You will remember that these trees are in the edge of a tongue of woods that juts out into a pasture and on the east side of it.

As I strolled along here, cocking my ear for the voice of the hound, the warmth of the nooning sun and the sheltered place invited me to a smoke and as I smoked my eyes were attracted by the peculiarity of these evergreens which the nakedness of the deciduous trees made the more conspicuous. The leaves and cones were so different from anything I had ever seen that I took home a branch to my wife who found out what it was by consulting Wood's botany.

Several years later when Mr. Pringle sent a man to me to have the trees shown him that he might get a section of one for the New Orleans exposition, we went to the owner of the trees for permission to cut one; I found that he was acquainted with them and knew just what they were,

"A cross betwixt the Norway pine an' a spreuce."

He seemed interested in the trees and quite pleased to have them noticed. He bore the very masculine name of Heman Bull. He is dead now and the land where the trees stand belongs to his son Peter. It is not worth clearing, but that is no sign that it will not be stripped though I think Peter would be inclined to preserve the pines if he knew their rarity. This, a tree lover who often sees him, has promised to inform him of. They were flourishing a year ago when I went to them, in the dark, probably for the last time, to guide a gentle man who wished to see them.

A boreal tree, *Pinus banksiana* covers hundreds of square miles in areas of sandy sterile soil in Michigan and Saskatchewan, but in northern New England is at the very edge of its range. It is a relict from an earlier, colder era, when mastodon roamed Vermont's forests and beluga whales swam in the Champlain Sea. In Vermont, it is known historically from the Bull farm grove, a solitary tree on the John Brooks farm in Starksboro (near the upper reaches of the watershed), and another single tree in Fairfax. More recently, a jack pine was found near Wolcott. This is the only surviving site known.

If he had been a trained botanist, Robinson would no doubt have slapped a representative sprig of the rare pine in a plant press to vouch for his discovery. Instead, he noted the event in his journal, and may perhaps have sketched the strange tree. In some way, the tree remained undiscovered. Its occurrence was not made known to the scientific world until 1880, when, visited by the naturalists Edwin and Charles Faxon, Robinson went on a day-long botanical outing, at the end of which he showed the brothers the grove he had discovered. The Faxons did take voucher specimens of the tree, establishing the scientific validity of its existence. Before this, the grove had been like the proverbial tree falling in an uninhabited forest; it made no sound until recorded on a herbarium sheet.

About a year later, an associate of C. E. Faxon's published an article on the forest trees of the eastern Adirondack region in which he noted some curious folk superstitions regarding the gray pine. Faxon passed along the information to Robinson in a letter:

> There the Banksian pine is regarded as "the unlucky tree," as a sort of Upas, poisonous to man or beast. It is even considered dangerous to pass within ten feet of the tree, and more so to women than to men. When one is discovered on a farm all the ills which have befallen the owner or his cattle are attributed to it, and the tree must be destroyed, not by cutting down, as no one would venture to do that, but by encircling it with a fire of brush or wood—like a witch or heretic. The more observant people consider it to be a cross between a pine and a spruce. I don't know how it strikes you, but it seemed to me one of the most curious bits of local superstition relating to plants that I have heard in a long time, and I am glad to see it recorded. How much of this thing is there left among our people do you know? I had supposed there was very little. Of course the people of Vermont are supposed to be too enlightened to foster such fancies. At any rate, I hope the owner of *our* grove knows nothing of the notion, for it would be a great pity to see these beautiful specimens of this tree sacrificed.

This belief most likely originated with a group of people with whom Robinson was intimately familiar—French-Canadians. The *habitant* believed that wherever the *Cyprès,* as he called it, took possession of the soil, it made the region sterile. (This was an astute observation of the habitat of the tree, but it ascribed causality in the wrong direction.) So powerful was its sterilizing influence that it was considered positively dangerous for a pregnant woman to walk near a jack pine. This belief was still quite strong in Quebec in the 1920s, and may be so in some rural areas today.

If Robinson found any vestige of superstition surrounding the gray pine among his "enlightened" neighbors, his journals and notes do not record it. He did record the Adirondack belief in the way he knew best, by weaving it into his fiction. It first appeared in a tale told by Uncle Lisha Peggs in the book *Uncle Lisha's Shop,* which was published in 1887. According to

Uncle Lisha, Noah Chase once encouraged Amos Jones to go along with him to club starving deer that were yarding up nearby on top of the crusted snow. Finding about twenty deer gathered around a small conifer, Noah clubbed ten of the animals and cut their throats. Amos begged Noah to stop slaughtering the helpless deer, but Noah merely laughed, and, in the most obscene act of all, clubbed a fawn-laden doe. Amos noticed that his companion was standing under an "unlucky tree," and warned Noah that something terrible would befall him. The blood-crazed Noah laughed again, calling Amos a "sup'stitious chicken-hearted ol' granny," and proceded to slit the doe's throat. Unable to stand any more of such cruelty, Amos headed home, leaving Noah to pursue a weakened buck. Chasing the buck, Noah caught the toe of his snowshoe on a fallen tree, and fell, breaking his leg and knocking himself unconscious. When he finally awoke and began to crawl home, he was haunted by an apparition of the fawn-laden doe. Pursued by this ghost and a pack of howling wolves until he could go no further, he collapsed in a clearing, yelling for help.

Though help arrived and Noah had his leg set by a doctor, his troubles were not over. He remained ill for three months, during which time he had to endure the incessant imaginary howling of the wolves and the vision of the butchered doe. This was but the beginning of his penance: Although his condition improved a little, Noah's son was killed by Indians out West, his older daughter ran away with a worthless drunk, and his other daughter married an Irishman! Noah himself eventually fell ill with consumption, and after suffering ten years, he died. Amos's prophesy is surely borne out, whether due to the unlucky tree or Nature wreaking revenge on a particularly ruthless enemy. When Lisha finishes the tale, one of the listeners, Joseph Hill, remarks that he doesn't know what an unlucky tree is, and Uncle Lisha explains:

> "Wal, what some calls an onlucky tree, an' thinks is, is a sca'se kind of a tree, half way 'twixt a cat spruce an' a pitch pine. The leaves is longer 'n a spruce 'n' shorter 'n a pine, an' the branches grow scraggider 'n any spruce. They hain't no size—never seen one more 'n ten inches 't the butt. They hain't no good, 'n' I d' knows they be any hurt, but some folks think they be, an' you couldn't get 'em to go a-nigh one for nuthin'."

Robinson gives his own impressions of the strange pine's powers in "The Gray Pine," a short story published posthumously in 1905. The story is actually a tale within a tale; the first part tells of a trip he took to the Adirondacks to hunt deer, having "been born where no large game exists." (Indeed, the Champlain Valley was in Robinson's day practically devoid of deer.) He joins a small hunting party guided by Uncle Harvey Hale, who positions each of the hunters along likely deer trails leading toward a river. After about an hour of fruitless observation, Robinson, as the nameless narrator, is forced to take shelter from a pelting snowstorm. He chooses a solitary evergreen, which he describes thus:

> The sheltering tree, which at first I had taken for a spruce, I now noticed was of a kind that I had never seen. It seemed to be, if such a thing were possible, a hybrid of the pitch pine and one of the spruces; its leaves too short for a pine, too long for a spruce, and wearing not the healthy, lusty dark green of either, but a hue of unwholesome gray. Though evidently old, it was low and stunted, as though it could draw no suitable nourishment from a soil that fostered other trees. The long branches writhed out in snaky curves from the lichen-scabbed trunk, and toward the ends were clasped by pairs of hooked cones like the warty claws of some unclean bird, and they hissed, rather than sang, as do the branches of the evergreen to the stroke of the wind. The bare earth about its roots showed no undergrowth of flowering woodland plants, but only some frostbitten fungus, black and foul with decay. A strange, uncanny tree, and it may have been a fancy begotten of storm and solitude, but I began to feel as if some unholy spell were creeping over me.

During a lull in the storm, a big buck comes toward him, chased by hounds, but as he shoots, the deer veers away and his shot misses. Uncle Harvey arrives and immediately discerns the cause of the narrator's unfortunate shot: "It's no wonder 't ye missed! It's more a wonder 't yer gun didn't bust er suthin' an' kill yer! Why, man alive, that 'ere's an onlucky tree! Come 'way from it!" Asked about the unlucky tree by the curious narrator, Harvey briefly relates a number of stories about misfortunes caused by the gray pine, but holds off from telling the "wust story." The next day, "a stormy one of sleet and snow and wild wind that no one who need not would go abroad in," finds Harvey and the narrator huddled around a stove sipping "pennyr'yal tea," a perfect opportunity to hear the story. What Harvey tells is pure melodrama, but its tragic ending is unequivocably linked to another gray pine.

It must have been as much of a delight to Robinson to discover this folk belief as it was to discover the little grove of gray pine on that September day in 1876. It hearkened back to a time when the landscape was more of a living thing, when the fates of trees and people were tangled together, an animistic era more reminiscent of the Abenaki than the European settler. That he recorded the superstition fictionally not once but twice testifies to the power it held for him. "Place" called strongly to Robinson, such that throughout his life he returned continually to the spots that enriched him. His journals record his perennial returns to the "Slab Hole" on the south shore of Hawkins Bay to hunt arrowheads, to South Slang in April to shoot pickerel, and in May to lower Lewis Creek to fish for black bass. After he became totally blind in 1893, these excursions were more often made in his mind than in the field. But the blind author was still able to make some voyages in the dark to those places that he immortalized in his stories. In the mid 1890s Robinson guided a friend directly to the spot on the Bull farm where he had discovered the gray pine twenty years before. Like one of the salmon that once ran in Sungahneetook, Robinson homed in on the little grove, detecting not by chemistry, but by the lay of the land he knew so well.

The Bull farm jack pines were still flourishing in 1902, when Robinson's wife and daughter Anna went, on "a perfect fine morning" in June, to see them. They asked Almer Bull, Heman's son, where the colony was, and he showed them himself. "I've been hoein' an' got the garden pretty well cleaned aout and wanter rest. I'll hop onter a horse an' go an' show ye jis' where they be," Anna quotes Almer in her journal, recording a bit of dialect. Finding the spot, Bull pulled up two small seedling trees and gave them to the Robinsons, and they planted them behind the house among a white pine plantation. In 1909, Cyrus Pringle visited the Bull farm grove, and collected specimens, but after that, the gray pines again ceased to exist. No botanists visited them, and there are no local traditions concerning the trees. If you inquire about "strange-looking pines" in the neighborhood, you will hear all sorts of stories about queer conifers, but none are gray pines. As far as the Brooks farm tree or trees, the same is true. I have talked to men who have hunted, trapped, sugared, and logged the woods there for forty years, and they know of no such trees. Wayne Hill, who has stewarded the farm that once included John Brooks's land for all the years of his long life, told me that he knew every tree on the property, and there were no gray pines. After a dozen fruitless searches of the woods in both localities, I tried dowsing for the old groves, with no luck. I even corresponded with a gifted dowser who "map-dowsed" them from his home in New Jersey. He learned that the Monkton colony was no longer extant. His pendulum did tell him that the nearest stand "of around 45 trees is very close to 4 miles almost due west of Hyde Park." So a man from New Jersey who has never been to Monkton knows that the old grove no longer lives and "feels" another living grove from a map, while no one within a mile of either the Bull or Brooks farm sites has ever heard of the trees. Like Frost's farmer's fallen star, they are invisible to those who would seem most familiar with them, visible and vital to the poet or soothsayer from afar.

Even though the trees seem to be gone now, one question remains: Why did they come to grow just where they did? As certainly as white cedar has an affinity for limy soils, jack pine "likes" sandy, sterile soil. In Maine, it grows primarily on granite; in New Hampshire the species is found only on Welch Mountain, a sterile granite mass south of the syenitic Franconia Range; in New York, it clings to the silicious soils of the Adirondack massif. In Vermont, *Pinus banksiana* is no different. The Bull farm grove grew on thin soil above the "red sandrock"—Monkton quartzite—which was exposed in many places. The entire ridge running from Fuller Mountain to Robinson's "Ruey Hill" is striped with open ledge, and the places on the ridge that now grow a dry forest of chestnut oak once supported many jack pine. The Brooks farm holds an outlier cobble-hill of Cheshire quartzite, and the tree probably grew on this rock, which is composed almost exclusively of well-cemented and metamorphosed sands. On such ledges and thin soils the more dominant species of the region—birch, maple, beech, and so on—compete poorly with the tough xerophytic pine.

This helps explain the distribution of jack pine at the edge of its range, but it is insufficient. There are countless sites like the quartzite ledge spots on the Bull and Brooks farm where the tree was found—why doesn't it grow in these places? The entire set of environmental variables—soil moisture, soil pH, microclimate, slope aspect, history of the site, and so on—comes into play, but there is something about ecology that cannot be measured. Since its inception as a science, ecology has grappled with this unquantifiable entity. In 1929, for example, at the International Congress of Plant Sciences, botanists heard a paper on "Chance as an Element in Plant Geography." When all is said and done, chance seems to have the last laugh, in the form of flood, fire, hurricane, or other calamity. For thousands of years *Pinus banksiana* has tenuously kept the Lewis Creek watershed as its home, shedding its seed while shifting between quartzite ledges. Its persistence is due to a few hundred million years of evolution married to serendipity. That it survived in a few places embraced by Sungahneetook is at once as likely and unlikely as Rowland Robinson's coming upon the strange pine on a September fox hunt, or Louis Fuller encountering the Beers surveyor so that the pseudo-meteorite would be immortalized. It was as likely and unlikely to happen as looking up to see a falling star.

The philosophy that Rowland Robinson espoused was stated most succinctly in an essay he wrote called "In Search of Nothing": "If one stays beneath the star he was born under, watching and waiting, it may, at last, prove a lucky one." Though his thirst for the spiritual was quenched by daily communion with Nature, Robinson inevitably faced the increasingly common question as to whether his own shining earthly existence might end in the same manner as Frost's falling star, burned out and forgotten upon some stone pile. Raised in a family of firm Christian believers who also were intensely devoted to Spiritualism, Robinson had modest expectations about the hereafter. On March 9, 1878, between comments on early spring bird arrivals and observations of a red squirrel lapping maple sap, something moved him to speak about the hereafter:

> I trust that this world, wherein there is so much of *heaven* now, will form a part of the heaven hereafter. I can imagine no more blissful heaven than the oases we find in this earthly pilgrimage, the green spots where we halt and rest with our beloved in the fullness of content. Wherefore should not a man take such forethought concerning the future comfort of his soul? If I lead ever so godly a life, can my spirit walk serenely among the desolated hermits of its embodiment? I know it will vex it when it visits these scenes, where this my part, if it find not the familiar trees, etc. even as it does now to return and find cherished objects of nature destroyed.

Each person built his or her own heaven. Robinson's father's heaven had emancipated slaves, equality for women, and it lacked paupers. It was natural that Rowland Robinson would imagine his heaven, should there be one, to consist of beech woods and wild rice marshes.

In true form, Robinson's published thoughts about mortality are small and subtle, coming in brief doses between fictional evocations of pathos and humor. In them he is still the skeptic; a life after death is not flatly denied, but the suggestion seems always to be that, as intimated by the diary entry, the *present* is enough. Assessing years of unsuccessful searches by one of his own fictional treasure hunters, Robinson stated that his "bairied riches never done him no good, 'thaout it was in expectin' on 'em, which is abaout all the satisfaction any on us gits." Robinson's uncertainty of an "undiscovered country" is affirmed also in his nonfictional discussion of Old Jesse Ball, for whom the first landing above the mouth of Lewis Creek was named. Speaking to one of Jesse's sons by the grass-covered cellar hole of Jesse's house by the Creek, the son pointed to one of his father's old charcoal-making pits. "There's the last coal-pit my father burnt—on this airth. I do' know haow many he's burnt sence."

It was in considering the death of a dog, however, that Rowland Robinson may have most nearly revealed his estimation of immortality. It was written while he was bedridden with cancer, a few weeks before he died. In *Sam Lovel's Boy,* Sam Jr. is out one summer day hunting woodchuck with his father's dog, "Drive," and the dog leaves the hunting to rest in the shade of a lilac. When the boy returns, he finds the dog dead. "The mysterious essence of life that dwells in men and dogs, and dreams dreams, had departed forever to the happy hunting grounds, where perhaps dreams come true."

Robinson's own death—on a beautiful autumn day, the 15th of October, 1900—became a test of faith for those who knew him. The writer Julia C. R. Dorr had visited Robinson at Rokeby a few weeks before, then returned a week after his death to the East Room, where Robinson had been born and died. She composed this sonnet as her eulogy to him:

No shadow darkens the resplendent day!
  O Mother Nature, dost thou make no moan
  When he, thy son and lover, lieth prone
Breathless and silent? All the hills are gay
In pomp of gold and crimson, and the play
  Of Royal banners shining in the sun
  Proudly rejoicing as for victories won!
Hast thy great heart no need to weep or pray?
And Nature answered:—"Nay, I but rejoice!—
  I bid my vales be glad, and all my streams,
  I bid my mountains crown themselves with light,
And every late bird lift a joyful voice;
  For lo! at length the radiant morning gleams,
  And he who once was blind hath done with night."

When Mrs. Dorr visited Robinson's room, she found next to his bed the grooved board on which he wrote, the pencil slipped in between the paper and the board, just as he had left them. Mrs. Dorr recorded the last sentence, which rings with a touch of the numinous: "The lifting veil disclosed the last flash of blue plumage disappearing in the mist of budding leaves from behind the cloud of smoke that now hid my mark." Mrs. Dorr found what she had come to the East Room for—some confirmation that "Awahsoose's" spirit lived on. But a different account of Robinson's last written words was given by Robinson's son, Rowland Thomas. He said that on his last day his father, though in great pain, had continued writing until about 3 P.M., when he laid the writing board aside, saying that he felt tired. An hour later he was dead, his last written words being:

> That end with a dull thud on the pasture sward.

This ending sounds less poetic, yet it captures the realism that its author held to in life, and so is perhaps a more authentic symbol of Robinson's unresolved journey to the "undiscovered country."

CHAPTER 7

Eden Lost

HERE IS A little tributary to Lewis Creek that needs to be followed to its source, for it originates in a spot of land that nurtured the watershed's most devoted plant hunter. This tributary comes in from the north just below one of the Creek's few persistent islands, a mid-channel bar rank with alder and ostrich fern. The tributary loops northward around a wooded cobble-hill, giving a view of what in some years is a common corn field, but in others is an uncommonly beautiful sight—a field of timothy spilling down a gentle southern slope. It knows its full glory on hot June days before the first cut, when there is a stiff south wind blowing. Then the tight spires of graminous flowers sway and reel and bend deliciously, making the land look alive.

The tributary heads in three places: easterly, in fields that have been for generations, and still are, farmed by Baldwins; the middle fork drains one of those classic Champlain Valley swamplands, elongated north to south and harboring the hulks of beaver-drowned trees; and the westerly fork joins the tributary at the south end of this swamp, coming from some seeps on the farm where Cyrus Guernsey Pringle was born.

The day before the summer solstice in 1867, Cyrus Pringle began to chronicle his relationship with plants:

> believing Horticulture to be one of the most innocent and ennobling avocations of man; recognizing it as the task assigned him by Infinite Wisdom and Goodness in the peace and delight of Eden, as almost his only legacy from Eden, and as one of the means to aid him in his Paradise Regained—his attainment of a better country than Eden—in that it offers the purest employment for his hands while his heart is employed under the Grace of God—I have sought to surround myself with fruits, to find in Horticulture employment for my hands, recreation for my impaired body, and relaxation and diversion for my mind.

Cyrus Guernsey Pringle. Courtesy Pringle Herbarium, University of Vermont.

That plants and their cultivation seemed to the Charlotte farmer human beings' "only legacy from Eden" was likely due to the violence he had seen his fellow men wreak upon each other, and that was the cause of his "impaired body." Pringle had been drafted into the Union Army on July 13, 1863, but as a Quaker deeply devoted to the ethical doctrines of the Friends, he had refused to bear arms when he reported for service. An uncle offered to pay the $300 commutation fee that would have permitted some other man to soldier in his stead, but Pringle declined. He and two other young Vermont Quakers (Peter Dakin of Ferrisburg and Lindley Macomber of Grand Isle) were arrested and sent to Camp Vermont on Long Island in Boston Harbor. At first, an effort was made to induce Pringle to serve in the hospitals instead of on the field, and he debated the compromise for days before deciding that "no Friend, who is really such, desiring to keep himself clear of complicity with this system of war and to bear a perfect testimony against it, can lawfully perform service in the hospitals of the army in lieu of bearing arms."

The three Quakers were confined to the guard house, where they kept company with "the subjects of all misdemeanors, grave and small . . . those who deserted or attempted it; those who have insulted officers and those guilty of theft, fighting, drunkenness, etc." Pringle noted that many of the prisoners were involved in the recent New York City draft riots, and that their racism was apparent. They "exhibit this in foul and profane jeers heaped upon those unoffending men [the black conscripts] at every opportunity. In justice to the blacks I must say that they are superior to the whites in all their behavior."

On the 31st of August the three men were called out one by one to answer questions regarding their offenses. Peter Dakin, asked if he would die before submitting to military service, replied promptly but mildly, "Yes." Pringle could hardly fathom the nature of their imprisonment: "Here we are in prison in our own lands for no crimes, no offense to God nor man; we are here for obeying the commands of the Son of God and the influences of his Holy Spirit. I must look for patience in this dark day. I am troubled too much and perplexed."

Pringle, Dakin, and Macomber were then moved to another camp in Culpepper, Virginia, where their treatment was even more barbarous. They were forced to march with guns strapped to their backs, taunted all the while by nearby enlisted men. One day Pringle was asked to clean a gun—when he refused, two soldiers tied him to the ground with cords about his wrists and ankles. With his arms and legs staked out in the form of an "X," he was left to lie in the midday sun for hours, until so weak he could hardly walk or even think. A corporal came by and urged him to give up, threatening Pringle with death if he did not submit. Pringle's only reply—"It can but give me pain to be asked or required to do anything I believe to be wrong."

Through the intervention of concerned Friends, who prevailed upon Isaac Newton, the Commissioner of Agriculture, an audience was granted to the three Vermont Quakers with Secretary of War Stanton. Though at that time public discussion of the treatment of conscientious objectors was intense, Stanton only paroled them indefinitely, telling them that his oath of office stood in the way of granting them a discharge. At this point, Pringle's strength gave way and he was hospitalized. On November 6 Newton presented their case to President Lincoln at the White House. Lincoln, who had always been sympathetic to the pacifist ideals of the Society of Friends, insisted that the protesting Quakers be relieved of military duty and sent home. On the return journey Pringle became delirious, and it was weeks before his health began to return.

The four-month ordeal precipitated by Pringle's convictions only served to strengthen, not weaken, his resolve, but it had left a trace of melancholy. The world of men was fraught with brutality, violence, and injustice. It was no wonder that Pringle found refuge in the green world of plants. He concluded the "preamble" to his first horticultural notebook: "Alas! should I ever pawn for it [Horticulture] any more sacred interests."

Cyrus G. Pringle
Charlotte, Vermont
20th, 6th mo., 1867
Lat. 44° 17½' N
Long. 73° 11' W

On that day before the summer solstice the Charlotte Quaker swore his allegiance to God's green things; he fixed the moment in time and space. The coordinates given are those to the place where a kind, compassionate farmer began a lifelong love affair with plants.

The next pages of Pringle's journal record the layout of his Eden: between the house and the barn lay the fruit yard, with a dozen or so Tolman's Sweeting and eighty-three Rhode Island Greening apples flanked on the west by a plum orchard. Lombard Plums, Blue Magnum Plums, Peach Plums, Coe's Golden Drop—these and other varieties formed a court for the poultry and their coops. South of these lay the piggery and pear orchard. On the west side of the house lay the "GARDEN," spelled out at the top of Pringle's diagram in letters of climbing vines on a trellis. Behind the north wall of arbor vitae fronting on the road, there were gooseberries, grapes, strawberries, asparagus, and rhubarb, then a plot for vegetables, fringed by currants and a few cherry trees. The life history of individual trees, marked with letters on his garden diagram, is given briefly: "Of the barberry bushes 'M' is very old having been planted about the time of the formation of the garden. 'N' was a present to my mother from Chas. McNeil several years ago; and Uncle Mark planted 'K' a little later." The date of full bloom of dozens of fruits and flowers is noted for successive years, as is the time of ripening of fruits, dates of first frosts, and the arrival dates for birds. These written records show the beginning of a phenology that poured itself into Pringle's very blood. The behavior of the many cultivated plants on his farm became as well known to him as that of his neighbors at Sunday meeting.

Pringle kept another journal that more thoroughly documented his *science* rather than simply his observations. On the first day of the year 1869 he noted that he had begun to specialize in the selection of fruit varieties particularly well suited to the Champlain Valley. Noting the importance of meteorological records to the understanding of vegetation, he lamented that only one such complete and extended record had ever been kept for the region (Zadock Thompson's, at Burlington). Pringle proposed a "chain of posts" for making meteorological observations, and wrote to Joseph Henry, secretary of the Smithsonian Institution, to see if such a scheme might be funded. Henry thought it wise but had no funds, and so Pringle began the network in his own small way by encouraging friends in Grand Isle and across the lake in Plattsburgh to keep records. Pringle made himself a rain and snow gauge to add to the data secured from his thermometer.

That winter the novice horticulturist wrote to the *Country Gentleman* to correct remarks made by one of their authors regarding grape vines. By spring, he received a complimentary subscription to the horticultural magazine, in return for his letters to it. He recorded the instance in his journal, saying that "it is the first time I ever earned anything with my pen." Kept from attending college due to the death of his older brother, Pringle was almost entirely self-taught. He visited or corresponded with as many horticulturists as he could. From Paris, in February of 1869, he received M. Henry Lecoq's work on hybridization, and, while waiting at the mill at Scott Pond on Lewis Creek to have his wheat ground, Pringle learned to read French so that he might study Lecoq's book.

Reading Lecoq's chapter on "Fecundation," Pringle was led to wonder about the horticultural possibilities of the many indigenous plants about him. In the hedgerows along the East Charlotte roads were plants like the pigeon cherry *(Prunus pennsylvanicum),* red chokeberry *(Pyrus arbutifolia),* purple-flowering raspberry *(Rubus odorata),* and the exceedingly variable shadbush (*Amelanchier* spp.). Further afield, in cold mountain bogs, there was the whortleberry *(Vaccinium uliginosum* var. *alpinum),* which Pringle also thought might be improved from its wild state by careful selection. The whole vegetable world called to the novice horticulturist.

Early on in the horticultural journal, in faint purple ink unlike that of all of his other entries, is a quote from Télémagne: "*C'est dans la fleur qu'il faut préparer les fruits.*" This quote presaged Pringle's major contribution to horticulture. Breeders before him practiced selection, but left the essential act of cross-pollination to the whims of Providence. Two different potato varieties, for instance, might be set in adjacent rows, but the insect that brought fertilizing pollen to any of those plants may have come from a distant row of completely alien stock. To keep reliable records of parentage was impossible. "*C'est dans la fleur qu'il faut préparer les fruits*"—man had to intervene more immediately into the reproductive process, at the flower.

After attempting to hybridize grapes during the summer of 1868, Pringle observed an interesting occurrence. Some of the grapes on certain bunches ripened with the rest, but reached only half the usual size. More importantly, these smaller fruits contained only abortive seed. Pringle suspected that this was caused by a deficiency in either the quantity or quality of pollen received, which he believed to be the key to the size and fertility of fruit. Charles Darwin, in his *Variation of Animals and Plants Under Domestication* (1868), reinforced this view. Before the deeply drifted snow had melted, Pringle was out tending to his fruit trees, with his winter's reading of Lecoq and Darwin heavy on his mind. In mid April, as he sowed his hotbeds, placing soot on the snow drifts that lay where he wished to set them, he meditated on Télémagne's incantation. As he pruned the orchards, inspected mice damage to young fruit trees, and as he planted early potatoes, he anticipated his new role as pollinator. To "prepare the fruit," he had to be at the flowers at just the right moment.

In mid May he worked with currants, crossing Versailles, White Grape, Magnum Borum, Fertile du Palluan, and Victoria in various combinations. The weather was consistently rainy, which slowed the flowering, but Pringle was content, for it also reduced his competition. "The insects are not out to interfere with my designs," he said. He observed closely the timing of flowering, noting how the blossoms unfolded by twos or threes, beginning with those closest to the woody stem. Selecting the heartiest flowers on the best branches on the most vigorous bushes, Pringle quickly removed all the other blossoms. Recognizing any flowers open from the previous day as already fertilized, he pinched these off also. Using small forceps, or the point of a pin, he picked the stamens out of the unfolding flower, then applied the adherent pollen to the stigma of flowers on the variety he desired to cross. Only skillful, disciplined hands could effect such a delicate operation without doing serious damage to the barely opened flower. Pringle once recorded his own unique qualifications as hybridist by way of transcribing a quote from Charles Darwin regarding plant breeders: "Indomitable patience, the finest powers of discrimination, and sound judgement must be exercised during many years."

He worked similar operations on plums, pears, apples, cherries, and gooseberries, then later in the summer got an opportunity to work on crossing tulips. These last required the protection of paper or muslin caps, which were drawn over the opening flower and tied lightly around the stem below. Pollen from ripe stamens was applied on successive days, each time re-covering the impregnated flower to keep it from receiving "stray" pollen. Pringle eventually employed this method with a wide variety of fruits, vegetables, and ornamentals, since not only did the little sacks (called *buxelles* by the French) protect against random pollination, but after fertilization they served to protect the developing fruit from insects and birds, and to preserve the seed when it fell.

Though it may have come as a shock some years later to his ex-wife, Pringle's obsession with plants eventually brought him some financial reward. In 1870, along with other hybridization experiments, Pringle attempted to cross a number of potato varieties with the Early Rose, the first promising commercial variety produced in America. The Early Rose was the result of the painstaking work of other horticulturists: In 1843, the Reverend Chauncey Goodrich of Utica, New York, had imported some seed potatoes from Chili at considerable expense. During the next twenty years he raised not less than 16,000 seedlings from these South American tubers. In 1861, Albert Bresee of Hubbardton, Vermont, produced the Early Rose from a naturally fertilized seed ball of one of Goodrich's varieties—the Garnet Chili. With its light red flesh and a surface as smooth as an egg, the Early Rose was a potato for everyone, aristocrats and hill farmers alike. For his pains (but more for the unrecompensed pains of Reverend Goodrich), Albert Bresee received from a New York seed company, B. K. Bliss and Son, the astounding sum of one thousand dollars per pound of true seed. It was the height of the "Potato Mania," the wild speculation about the unprepossessing tuber that was reminiscent of the tulip madness in Holland in the seventeenth century.

Pringle appreciated the work of his predecessors, but believed that they had not proceeded far enough in bringing out the full extent of variation contained in the genes of the South American potatoes. He predicted that "the slow gains of former years will compare but poorly with those which must follow from an intelligent and extended system of development employing hybridization, that swift and fertile means of variation." But although Goodrich and Bresee were content to leave cross-pollination to the whims of Providence, Pringle was not. His experience showed him that careful, manual fertilization of flowers was imperative.

Pringle was determined to carry the course of breeding one remove farther from the wild state of the original plant. He had learned that in the Vermont climate, aborting the stamens (the male reproductive part) of the Early Rose rendered it a pistillate or female plant, incapable of producing its own seed. To effect cross-fertilization, he selected the Excelsior, a white potato of good quality and fine flavor, as the staminate or male plant. Instead of shaking the flowers of the Excelsior over those of the Early Rose, Pringle directly applied pollen from the bursting Excelsior stamens onto the Early Rose stigmas. A few days later he delighted to see a number of seed balls forming.

From this cross, about a hundred new varieties were raised. One by one these were culled by their creator as he reviewed them for color, form, texture, and eating quality. His "No. 6" was chosen to be most ideal, and he distributed seed of the new variety to acquaintances, who had great success with it. In January 1874, B. K. Bliss negotiated with Pringle to introduce "No. 6" to the public as the "Snowflake" potato. In return for entire control of the stock, the seed house was to advertise it, pay freight, and return to its originator 55% of the sales.

The work of artificial selection presupposes human control of nature, yet Pringle seemed ever aware that humanity was part of nature. The techniques of selection and hybridization Pringle recognized as a new art, but not one without earlier, equally skilled artisans:

> Among the flowers under my window this bright June morning are a myriad of hybridists at work, and thus they have plied their art ever since flowers bloomed and sun shone. The hummingbird, lightly poised in air above each flower, his plumage glistening with gold and green, and his long beak dusky with pollen, which unwittingly he carries from flower to flower; the blustering bumble-bee, in yellow and black, stirring up the precious fertilizing dust, gaudy butterflies furnished with the most delicate brushes conceivable, and flies and other insects of every name and degree.

These "busy operators," not chance, acted upon plants to cause variation. Another agent was the farmer's treatment of the soil that nourished domesticated plants. In 1871, in an address to a group of farmers assembled in Brandon, Pringle made this point by way of an example:

> The inferior and degenerate variety of wheat on a cold, thin and poor soil only yields five or ten bushels per acre. The most productive variety we grow could not do much better in such conditions. But by underdraining and deep tillage and high manuring, make that same soil warm and fruitful, and again sow the same varieties. The inferior sort may now yield twenty bushels; while the other would likely to give forty.

Pringle felt that the degeneration and decay of many cultivated varieties was due not to careless breeding, but to planting of inferior seed and indifferent cultivation. Every farmer, he said, needed to become a breeder of plants, in the sense that he choose for planting the best and most viable seed. After that, he must carefully steward the plants and the soil from which they sprang.

On his farm in East Charlotte, Pringle was recapitulating in a few years time much of the horticultural observation which had gone on in previous centuries. That his primary reference was Darwin's *Variation* was fitting. Published in 1868, this two-volume work was essentially the data from which Darwin prepared the abbreviated *On the Origin of Species by Means of Natural Selection*. If the *Origin* was, as Darwin put it, "one long argument from the beginning to the end," then *Variation* was an even longer argument. But Cyrus Pringle's thoughts on the most revolutionary idea of his century go largely unrecorded in his horticultural journal. Though he quoted Darwin often, he rarely commented on these passages. On February 18, 1870, Pringle wrote: "From Mr. Darwin's discussion of the subject, it would appear that acclimatization is effected not by modifying the habit of an individual plant, or, which is the same thing, of its successive generations produced by division; but by the selection of new characters appearing in its seedlings." Here was the germ of Darwin's theory, and Pringle had not a word of comment upon it. Just as later in life he would be absorbed completely by plant collecting, Pringle in the decade after his Civil War experience was totally consumed by his horticulture. He was so much occupied with the logistics of hybridization that theoretical issues like that of speciation, though it was hotly debated by both scientists and the lay public, seemed to elude him. And it may have been due to this obsession with plants that something else eluded him, something that left another melancholy in its wake.

During that period of his life when he was farmer-horticulturist, the Pringle family consisted of Cyrus, his mother, his younger brother, and his wife and daughter. Almira Greene, a schoolteacher from Starksboro, was a

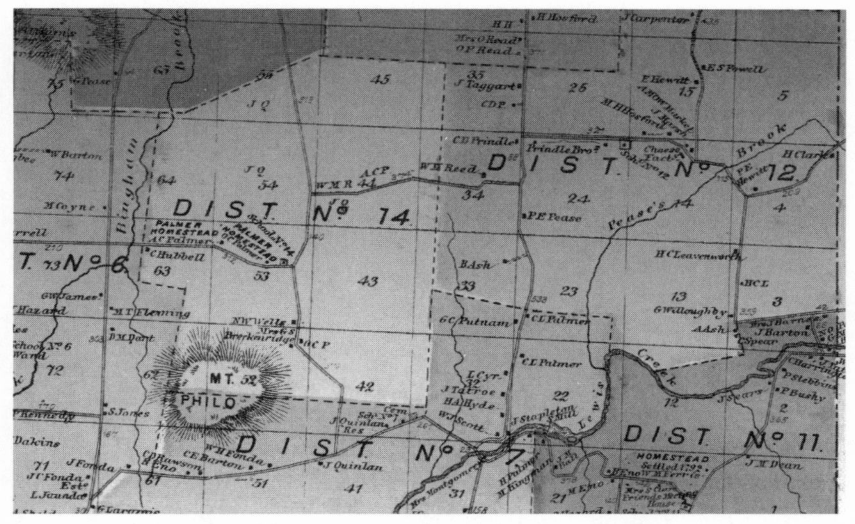

From F. W. Beers, *Atlas of Chittenden County, Vermont* (New York: F. W. Beers, A. D. Ellis, and G. G. Soule, 1869).

skilled speaker at Friends meetings when Cyrus had first become attracted to her eloquence and idealism. They were married in February 1863, and in October 1864, a daughter, Annie, was born. Almira, completely devoted to Quaker principles, probably encouraged her husband's conscientious objection as supportively as she could. She was unable though to support him in his devotion to plants. Not only were his experiments in horticulture largely unrewarded financially, but he had begun to spend some of their small income on herbarium specimens that he purchased from other collectors. This was irksome enough to her that early in 1872 the Pringles separated, Almira taking with her their only child.

Many explanations have been given for the couple's separation. At the time, people were told that Mrs. Pringle was zealous to engage in evangelistic work, and wished for Cyrus to participate, but he believed that he had neither the taste nor talent for it. Some said that she was in poor health, and she was persuaded that it would be better for her to live with her own mother rather than her husband's mother. Others whispered that it was simply a conflict between two strong Yankee women—mother and daughter-in-law—and that it was inevitable that the mother-in-law prevailed. But Cyrus himself confided to his best friend, George Davenport, that it was his pursuit of botany that wrecked the marriage, and years later, as a lonely old man, Pringle intimated to a former student that his wife resented his spending money on plant collecting.

The differences between husband and wife seem to have been irreconcilable, but what of the young daughter, who was only eight when her parents separated? No journal entries record the joy that must have greeted the newborn daughter from a father who only a year before had returned from three months of incarceration; no diary speaks of a loving father's many hours spent in the company of his only child; no photographs capture the tenderness that must have dwelled there in the East Charlotte homestead. The separation was permanent: On October 16, 1877, a formal divorce was obtained, with the wife receiving $2,000 alimony and the custody of Annie. Pringle's ex-wife was embittered and determined to keep Cyrus from seeing his daughter again. A man who was steeped in heredity, connected tenaciously to a vast history of Earth relationships, lost his only hereditary connection to the future.

It was sheer chance that years later brought Cyrus in touch with his daughter. On Christmas day in 1883 Davenport wrote to Pringle with some startling news:

> My sister is matron of the Consumptives Home in Dorchester, [Massachusetts] and while out here Sunday the children who were showing her the photographs [of us] pointed you out by name. . . . 'Mr. Pringle,' I heard her say. 'There is a lady by the name of Prindle and her daughter coming from New York to take charge of the Orphan's Home out to our place *who has been engaged in work for Fallen Women in Buffalo*.' At once my suspicions were aroused and a few questions satisfied me that here was your lost daughter coming almost within my reach. . . . I wonder if I was to be an instrument for bringing her and you into communication and restoring to you her love. I shall watch for such an opportunity.

Pringle replied that he had heard that his daughter was going to Boston, and had hoped that he might "through some good friend . . . put a letter into her hand," but had no idea how he would do so. He implored Davenport:

> I beg of you not to let [this avenue of communication] . . . be lost or closed by reasons of suspicions aroused in her mother. How I wish you might meet my daughter and disabuse her mind of the false notions concerning me, which have been assiduously implanted in it. . . .
>
> Now I am glad that I told you so freely and fully my feelings toward my poor daughter. I have now little hope of enjoying her society, while her mother lives. It may be best that we should not meet; I want her to know that I have cherished a father's love and that there are good people, who love me, as they understand me.

Within the year a reunion of sorts was arranged, "thrilling me with strange emotions. Will patience really have its reward?" Pringle wrote to Davenport. The mother continued to conspire against reconciliation—four months later, Annie visited Davenport, and Pringle told his friend that "my daughter (alas, how little mine!). . . is not yet more than half disenthralled from the influence hostile to me, under which she has been reared." The hostility only increased with time, and for some reason Annie never made her separate peace with her father. In 1900, when he made out his will, Pringle left his daughter $100, which, he wrote, "for reasons satisfactory to me, is all I desire to give her."

At the age of fifty, romance entered Pringle's life briefly in the person of Miss Kate Furbish, a botanist from Maine. (She was immortalized in 1976 when a rare plant discovered by her a century before—*Pedicularis furbishiae,* the Furbish lousewort—was relocated at its original site, halting construction of the Dickey–Lincoln dam project on the Saint John River.) The two botanists corresponded regularly about their common passion—botany—until they grew bold about each other. A week after his fiftieth birthday, Pringle intimated to Davenport: "Last evening came a letter from Miss Furbish; she regrets that she cannot see how I look—that I did not show myself in the view of a mesquite forest in *Garden and Forest* [a botanical magazine] . . . Miss Furbish and I . . . are getting more than ever before interested in each other. She is wishing herself young again that she might have strength; I feel like wishing ourselves both younger for other reasons—Don't tell!" But New England and its botany grew more faint with each passing year, as Pringle went deeper into Mexico and stayed longer. Freed from the ties of a family, he had become a vagabond for plants, a freedom not without its benefits for a man who had largely expatriated himself to Nature already. He once confessed to Asa Gray while in the Santa Rita Mountains of Arizona, "Exile in these wilds has one compensation not lightly prized by me; I don't hear the names of a presidential candidate once a week."

His own name, however, he still longed to hear spoken by someone who could love him. On his Mexican forays, he asked Davenport for news of Miss Furbish, but the relationship withered and died. Then, in 1894, his mother died. The loss sent him back to the field for refuge, but upon his return to Charlotte each time he could think only of his mother. To Davenport he lamented, "I am desolate here and half-sick from mourning for my precious mother, as I have put away her things to make room for the stranger who has come to take care of the house in my absence." Davenport sent photographs that he had made of Pringle's mother, and Pringle thanked him, saying, "I cannot hope to ever find one to comfort and love me in her stead."

The melancholy that haunted Cyrus Pringle was both the cause and effect of his botanical wanderings. For thirty years he tramped the continent in search of plants: from the rain forests of the Pacific Northwest to the "Cardon" (giant cactus) gardens of Baja; from the deserts of Sonora and Chihuahua to the great barrancas of the Sierra Madre. Even his Vermont work had taken him far from the paths of most men—Smuggler's Notch, the summit of Camel's Hump, the cliffs at Lake Willoughby. Pringle's botanical wanderlust brought him to these wild places because he loved plants, but he couldn't seek them out as fervently as he did unless he bore at least some disaffection toward society. The tens of thousands of miles he traveled in Mexico, away from his East Charlotte home most of the year, only added to his estrangement. In Mexico he contended with desert heat, raiding Apaches, peasant banditos, malaria, and a host of other difficulties that made plant collecting a hazardous and solitary occupation. His friends often advised him to give it up. To Davenport, he replied: "Do you despair, and fear that I will cling to my infatuation for botanical collection as long as life lasts? Well, what else am I good for?"

Like the soil and climate bringing a certain expression to the genetic characteristics of a selected plant, Pringle's environment had conspired to express in his life an inspired but lonely peregrination among plants. On May 6, 1838, on the farm drained by the three-pronged tributary of Lewis Creek, there had come into the world a living being formed yet unformed, preadapted perhaps to introspection, predisposed to an infatuation for plants, a likely candidate to become a botanical collector. But these predispositions came together in someone born at 44° 17½' North Latitude, 73°11' West Longitude, a man born into a watershed harboring wonderful botanical discoveries, to one who would as a boy get to tramp over rich limestone cobbles, to cold cedar bogs and bare quartzite cliffs. As genes and habitat conspired to produce a particular form of red oak, or trout lily, so did they conspire in Cyrus Pringle. The land bore this botanist, and it sustained him: "I have sometimes thought [that] if I were shut up in a prison yard, I could yet within its narrow limits find enough of God's handiwork to study to make me happy. This I have thought as I ranged over my wide, exhaustless fields, it is true; but I have in my day been shut up in prison quarters, without a bit of yard, and under circumstances that admitted of but little hope, and I am sure I was not even then miserable." But Pringle had a capacity for human love as large as his love for plants, and an even larger desire *to be* loved. His success as a botanical collector grew each year, yet he had no one with whom to share his fulfillment.

In 1899, Frank and Edith Estey of Charlotte came to the Pringle homestead to serve as caretakers, so that they could keep the old farm running in his absence. Pringle usually took with him one or more young farmhands to serve as assistants—in changing plant dryers and actual collecting in the field—and two of Frank Estey's brothers had served in this capacity (Bert Estey to Cuernavaca in 1898 and Bert and Walter Estey in 1899—their other brother Wallace accompanied Pringle in 1900). On January 17, Pringle, Bert, and Walter Estey left Charlotte by train for El Paso, Pringle's usual entry point to Mexico. Two days later Pringle wrote Frank from Chicago, telling him of their adventures there:

> We . . . have been five miles or so to the north to see the big Ferris Wheel and Lincoln Park with its greenhouses and its cages of wild animals. The lions roared tremendously and shook the house, as though for our benefit. Then we took the elevated railroad for Jackson Park . . . (city all the way and far beyond) to see the site of the last World's Fair and the Field Columbian Museum with its wonderful collections—we could not see much in the two hours we had there, but we glanced at a thousand strange things.

In this and other letters home, Pringle inquired after and dispensed advice on farm matters: suggestions on where to prune in the orchard; when the hay should be cut; where to pasture newborn calves, and how to keep mosquitoes from getting into the house. He asked anxiously about bills to be paid, incoming correspondence, the health of friends and relatives, and his cat, Tony. Financial matters received the greatest attention:

> Well, I can tell you of our having made a good beginning here [near Mexico City]. We have 25 plants for sets—$125.00 worth in ten days. . . . Many a bush or tree yields us $5. Last night Walter and I came in from a two days trip over the mountains with $40.00 worth . . .
>
> How good to have some money in the bank again!
>
> Please pay Mr. Patterson's bill by money order as soon as convenient. . . .
>
> You must draw money from the bank to make up your monthly pay—How far does the milk money go? . . .

> The Mexicans paid me $332.00, silver, for the plants I brought from home, and money will be coming for logs, so I guess we shall not have to draw on you for a time.

The profit margin in plant collecting was slim, even with free passes for rail travel, unpaid assistants, and, in Pringle's case, a spartan lifestyle. When he had first started collecting in Mexico, he had an official appointment as Botanical Collector for Harvard University's Gray Herbarium, which gave him a maximum salary of $800. In 1892, though, after the death of Sereno Watson (Asa Gray's successor), this stipend was discontinued. Pringle's finances were then so tight that he had thought of selling his herbarium to the American Museum of Natural History, but Asa Gray's widow intervened, lending her husband's favorite collector a thousand dollars so that his herbarium would remain intact and so he could continue exploration. Years later, Mrs. Gray burned the notes for this loan in Pringle's presence, saying that she felt she could gladly give this money for science, in memory of her husband.

Even with such generosity, it was always a struggle to stay in the black. There were farm expenses, the hired man's salary, printer's bills, shipping expenses, and herbarium paper to be bought. These debits had to be paid out of an income entirely at the mercy of the elements: If there was a drought in a region where Pringle was working, no plant material could be gathered, and there was no income. "Sets"—a collection of single specimens of each species collected on a particular excursion—of Pringle's Mexican plants sold for $10 to $20 ($30 in later years), and he usually distributed more than fifty sets to public and private herbaria, so if the weather favored him, he could expect about a thousand dollars from each expedition. But droughts were not infrequent, and neither were epidemics of malaria or other diseases that prevented Pringle from ranging widely in search of plants. There were a number of times when the Charlotte farmer made the 2000-mile trip only to find that he must return empty-handed.

By early June of 1899, Pringle and the two Estey brothers had exhausted that season's plants in the Mexico City region, and they returned home. Two months later Pringle was back in Mexico City, this time with Frank Estey. Whether Frank had asked to go because of the reports of exotic people and places from his brothers or whether Pringle asked Frank to go along is unclear. From his correspondence, it is obvious that Pringle wished to become better acquainted with his hired man. This was hardly surprising, since Pringle left in Estey's charge the farm that had sustained three generations of Pringle's family.

Upon their arrival in Mexico City, the two men took up quarters in the Hotel Buena Vista in the west end of the city. This was then the modern, growing section of the city, with open squares and broad, clean streets and boulevards, adorned with monuments and statues, sometimes bordered with trees, full of white sunshine and flanked with elegant mansions. Pringle contrasted it with "the East End of the city, the ancient Mexico of the Viceroys, crowded, dingy and dirty." The Buena Vista was the perfect base for their collecting operations, located across the street from the railroad station of the Mexican Central and Cuernavaca lines:

> A mile away in the Zocolo, or central plaza, I could take cars of the District and Valley Railroads, the former leading due south ten miles over wet meadows to Tlalpam, planted at the foot of the hills and close to the east edge of the lava beds; the latter bearing into the southwest, first over soft meadows, then over dry land and through the suburban villages of Tacubaya, Mixcoac, and San Angel to Tizapan, which is situated on the west edge of the lava beds, toward the base of the peak of Ajusco.

As had happened to so many of his young Vermont assistants unaccustomed to the Mexican climate and food, Frank became ill almost as soon as they were settled in Mexico City. For the first week, though he ventured out once or twice to collect plants in nearby locales, Pringle acted as nursemaid for his hired man, and quickly decided to send Frank home. Taking him as far as El Paso, Pringle realized that Frank was too weak to get home alone, and he accompanied him all the way to Charlotte. Leaving Frank safely in his wife's hands, Pringle and Bert Estey made hasty preparations and returned to Mexico the following day.

The first day of their trip back, they were crossing the Saint Clair River in Michigan by ferry, and Bert went up on deck to see the sights. Pringle, however, stayed below and wrote to Frank: "I can feel no fresh interest in them, but prefer to turn my thoughts to you. Much of the time throughout this long day I am thinking of the precious friend who was so lately with me. By the pain I feel today in my breast, as I travel far away from you, I know that I have left my heart behind with you." The next day they passed through Kansas City, and Pringle wrote to Frank: "It seems so hard that I cannot be near you, since my happiness and prosperity so greatly depends on you. You tell me to look forward to the days when we may live together and work together—I shall hope for such days and work to realize such happiness. But last month I did hope for the same things; and now the shadow of our recent disappointment oppresses me as well as a sense of the perils and toils that must be passed before we see those days." Two days later Pringle wrote from El Paso, telling Frank that he and Bert had taken their bath "where you *and* I took ours." There and at the Buena Vista in Mexico City, the two caretakers became lovers. In the midst of kindness and compassion—a young man tending Pringle's home and an old man tending the fevers of his hired man/botanical assistant—Pringle found someone to love.

Back in Mexico City, Pringle wrote that he was tired, since he had spent most of the previous month on trains. In spite of his enormous capacity for travel, for solitude, and for hard work, Pringle knew that he was tiring. He recognized that he was growing too old for constant travel, and he still had difficulties finding and keeping botanical assistants. He wondered aloud to Frank whether he should remarry and make shorter expeditions, or give up botanical exploration altogether. It seemed that he could only go forward if there were someone to love, and to love him: "Unless I can love those of my own (or some quite as dear whom I may call my own) to love with all my soul my remaining years will be desolate, and my end may be dreadful."

Pringle wrote often to Frank, and returned from the field each day hopeful of receiving a letter in return, but none came. The weather mirrored his soul; on September 21st, he wrote in his diary: "Today the sun scarcely appears over the Valley of Mexico, so dense and low are the clouds, but no rain falls." Two days later Pringle was out on the *pedregal,* the lava beds between Tizapan and Tlalpam, hunting for plants on the sharp ridges and in the deep crevasses of the black rock. It was a landscape as capable as any of taking the lonely collector's mind off his unrequited promises of love. It was a mythic landscape reputed to have been rife with dangerous men, wild cattle, and venomous reptiles; such myth and the rigors of travel there had kept most botanists away, so it was a rich and rewarding ground for Pringle. Plants—of mountain, valley, and plain—he had first collected in other states, he met with here, always in some inaccessible recess.

Coming home from Tlalpam after dark, Pringle met a thunderstorm, and rode home soaked to the skin. At the hotel though was a letter from Frank, which heartened him. The weather changed. In his diary, he noted: "The weather is lovely; even the summits of the great mountains become cleared of clouds, which would indicate that the rains are about over. The wind blows cool from the north." To Frank, he spoke little of the weather or of his plant collecting, but of his longing. "Tonight, as usual, . . . the one great wish of my heart is *to be at home with my precious Frank* to rest in his company."

A few weeks later he was with Frank, spending most of the winter at home in East Charlotte. Before Pringle left again for Mexico in late March, he had given Frank the power of attorney over his business affairs. As before, Pringle wrote to Frank often on the trip down, and once he had arrived in Mexico. This time, before settling in Mexico City to work, Pringle explored in the desert state of Coahuilla. The winter rest had given him new strength—he got into the fields most days around 7 A.M. and worked ten or eleven hours in 100° heat into the early evening, foregoing dinner, and bringing in "$25 worth of plants nearly every time." But Frank had not written him in nineteen days, and again he worried. Though he asked how the hay and feed were holding out on the farm, and whether the grass had started, his main question was whether Frank still cared for him.

Pringle and Bert had traveled to Jalapa, in the state of Vera Cruz, near the Gulf of Mexico, to work ground that Pringle had found fruitful the previous year. When they arrived they found that there was a smallpox epidemic, and though they both had been vaccinated against the disease, they were cautious, and hoped to leave the region quickly. Before they could get away, Bert came down with malarial fever, and Pringle again became nursemaid. Ten hours of fieldwork and evenings of doctoring his assistant could not fatigue Pringle enough to give him a restful night's sleep. Anxious about his future, he lay awake for an hour each night. Once he got up and spent the restless hour writing of his concerns to Frank.

> Now, do you think it strange that I should get very wide awake betimes in the night, when things are so hard with us? Or can you, after what I have before written of the pain it gives me to be separated from the one I love best, blame me for a longing to get home again to share in that happiness which you write of? But how much of a share of that happiness can I hope to enjoy? . . . You did make me happy last winter—almost happy enough. But can it be possible that you was [sic] beginning to forget me in two or three weeks after my leaving home? . . .
>
> When you speak of your love for our good home, and wonder how long you may enjoy it, you touch a tender place in my heart, my precious boy. When I left you at the end of March, you were all to me that my heart could desire. I am unwilling to fear that I shall find you on my return one bit less kind and true. Our pretty plan is on trial, and we are testing each other. . . . I have had so far a life of sorrow and hard toil. I do want to fill my remaining years with as much joy and peace as possible.

A week later Pringle started for home, and before leaving on the second Mexican trip of that year, he settled his fears for the future. He made out his last will and testament, leaving to Frank Estey the Pringle farm and all of his personal property. On his way south, he wrote: "Now I feel that I have something to live and work for—that there are joys to look forward to."

On this trip Pringle took with him as assistant Frank's youngest brother, Wallace, but even before they had reached Mexico, Pringle realized it was a mistake. Wallace complained about everything, and his negative attitude brought on problems. At customs inspection in Diaz, they were given a hard time, and a day after they got to Mexico City, Wallace contracted malaria. The two started back to Vermont, Pringle disappointed, but happy to be returning to Frank. The month they spent together only made it harder for Pringle to leave. From Las Vegas, he wrote back to Frank, his anguish so great that his pen, as well as his words, revealed it. In a postscript he said: "A year ago come tomorrow I was in this same vicinity, and with me was my dearest friend on earth. Now I sit alone and desolate. What is the outlook? Can you see any more joy for me ahead? Danger enough!"

The following months brought letters from Frank that allayed Pringle's fears. If questions of home and family seemed settled though, the question of his life's work was not. Pringle's 1900 will had designated Middlebury College as the recipient of his herbarium, library, and archives, but in 1901 he was in Washington, D.C., working at the National Herbarium with his friend and botanical associate Joseph Nelson Rose. Rose was plotting furiously to win a place for Pringle and his herbarium there. Ezra Brainerd was doing similar finagling with the Middlebury trustees. Pringle himself felt he belonged at Harvard, but this wealthiest institution, for which he had done the most, failed to offer him the security he needed. In June of 1902, when a horse-drawn wagon carried Pringle's life's work from the old brick farmhouse in East Charlotte, it headed north to Burlington, not south toward Middlebury, Cambridge, or Washington.

On that wagon were boxes of plants, some fifty thousand or more, gathered at first from the fields around the Pringle homestead, then along Lewis Creek, out toward its various headwaters—from Monkton Pond, Bristol Pond, and the furthest reaches cutting the phyllite knobs of the Front Range of the Green Mountains out in Starksboro. Beyond the watershed, within sight as the wagon headed north along Dorset Street, were the peaks of Camel's Hump and Mount Mansfield, whose alpine flora had first put wanderlust in the fledgling botanist's eyes. The vast Samalayuca dune fields of northern Chihuahua, the volcanic mountains and *pedregal* near Mexico City, the subtropical forests of Oaxaca—these and other exotic locales were frozen on the herbarium sheets aboard that wagon.

The appointment at the University of Vermont was as "Herbarium Keeper and Collector," paying Pringle a life annuity of $600 with an additional $350 annually for maintenance and expenses. Finally possessed of the institutional affiliation for which he had yearned for almost three decades, he was still not without remorse. To his friend and fellow Vermont botanist Willard Eggleston, he lamented: "My herbarium room now sounds hollow; and while I have been putting up plants to send abroad, I have had occasion to refer to my herbarium often—*Must* I follow it to Burlington?" He would follow it, but not just then. First he was off to work the mile-wide barranca near Guadalajara, following the zigzag trail down toward the foaming Rio Lerma, searching its cliffs for new plants. As he did so, his new home back in Burlington was being laid out for him in Williams Science Hall on the university campus. He wrote to Frank:

> If you go to see the chalk-marks, you may learn, what Prof. [Lewis R.] Jones writes me—that he and Mr. Votey have decided to include in the herbarium all the space south of the south wall and to make between the two walls a wide hallway with a storeroom on the east side and a room for me on the west side. That will make a handsome finished job, finished for all time. I am wondering if that room will not be my future home, and if my sweet and precious friend and I will take our naps there. Or must we part forever?

The following winter Pringle joined a number of Harvard botanists in Cuba to work on crossing sugar cane, then returned to spend the summer in Charlotte before leaving to roam the Sierra Madre. Gradually he distanced himself from the old homestead, spending long hours on his Vermont stays in the new herbarium. The hopes of his last years being spent in loving company with Frank faded, and more than ever, it was for his plants that he lived and worked. By the end of 1905, the herbarium was undeniably his home, as he kept a cot next to the table where he prepared his herbarium specimens. But Pringle found it hard to call it such: "As soon as I can bring it about I mean to start for Burlington—I can hardly say *home,* for I can hardly feel at home in the University. Yet it is all the home I have left now. . . . I am sure that I wish never again to enter the dooryard you mention nor to see the old home again. It hurts bad enough to remember it."

Two years later, Pringle still was not at home on the fifth floor of the Science Hall, even though he was surrounded by his intimates—plants he had brought back from Mexico. He admitted to Frank that he had not escaped his loneliness by going to Mexico again, where he had developed for his Mexican assistant Filemon Lozano an affection like the one he had for Frank. His only comfort now seemed to be memories of times past. His last letter to Frank closes with a wish for the future, but it is unmistakably a lament, not an expectation: "I have many times wished I might have received another visit from you before I left. But you came once, it was like old times. I could not ask you to come again, though I wanted to. But let us hope to meet many times yet, before one of us goes hence."

Four years later, the prince of botanical collectors went hence. In May of 1911, Pringle was in Burlington, planning a collecting trip to South America, when he took a long walk to visit a friend. The exertion of the walk brought on a case of pneumonia, and on the 25th of May, two weeks after his 73rd birthday, Pringle died, having nothing of his own but his quarters in the Science Hall and his plants. During his last delirious hours he talked of Mexico and of his happiness there.

Frank Estey, the man who in Pringle's last years represented love, family, and hope for the tired and traveled plant vagabond, was not one of the pallbearers at Pringle's funeral, but he came eventually to rest alongside the man who had once nursed him, and loved him so fervently. The two men are buried in the Morningside Cemetery in East Charlotte, about a mile from the Pringle homestead. The big iron gates to the cemetery always lie half open, and behind them the lilacs and white cedars seem good company for the dead as well as the living. Pringle's resting place is at the crest of the gentle slope that defines the cemetery. On the south side lies his mother, on the north, Frank Estey. The granite headstone that marks his grave ironically continues in death the anonymity that shrouded Pringle in life. The stone reads "Cyrus Guernsey Prindle 1838–1911"—the relatives who buried Cyrus never agreed with the way he rendered the family name, and spelled it their way, not his. But Prindle or Pringle, the gentle plant vagabond had regained his Eden.

CHAPTER 8

Windows

Here are three distinct successive periods of existence, and each of these is . . . a thing of indefinite duration . . . The result, therefore, of this physical inquiry is, that we can find no vestige of a beginning, no prospect of an end.
—James Hutton

RIVER IS A great exposer of things hidden, digging up the past as it makes a new present. It sweeps slowly across the landscape, cutting through soil and rock as inevitably as the wind and rain. At its most persistent, it cleaves the continent, opening up epochal memories engraved in stone. The Colorado River, for example, exposes the entire sedimentary span of Paleozoic time from Cambrian to Permian in its run to the sea. The Colorado strips country already laid bare to some extent not by water but by its absence, for the desert is only spotted with vegetation. There, rocks define the landscape; plants are an afterthought. Lewis Creek, however, cuts open a land of both abundant rain and past glaciation, where green plants and an assortment of glacial till, marine, and fluvial deposits shroud the rocks and the earth history encapsulated in them.

And yet there are windows, views into the geologic past that seem all the more precious given their uncommonness. Like its brethren west-flowing rivers to the north—the LaPlatte, Winooski, Lamoille, Mississquoi—Lewis Creek in its run from the Front Range down to Lake Champlain makes an incision through geologic time. It helps explain the earth events long past by sweeping them clean of obscuring trees and shrubs and Pleistocene cover. In some places the Creek still works on the rock; in others the naked rock lies high and dry, but in such a way that it is free of plant cover. To these ledges go those interested in the earth's history.

Of the rock exposed by Lewis Creek, none is as striking as the dark red to purplish rock that outcrops in the watershed from the line of "Overthrust Hills"—Pease Mountain, Mount Philo, Mount Fuller, and Shellhouse Mountain—east to the foot of the Green Mountain Front Range. This crimson ribbon of rock extends north to the Lamoille River and south almost to the Massachusetts line, but it is in Monkton that it is most thoroughly and frequently exposed. All rock formations, like plants and animals, have a "type locality," a place where they were first described thoroughly. The town of Monkton—unique in the United States in that it is the only town of that name—gives its name to the rock. "Monkton Quartzite" to today's geologists, to an earlier generation of non-scientists it was "red rock," and to the geologists of the nineteenth century it was the "Red Sandrock." Though geologists then agreed on the name, there was little else about the rock that they could agree on for nearly a century.

The Red Sandrock received its first introduction in print to the scientific public in 1842 via Ebenezer Emmons's *Geology of New York, Part III*. Along with his New York work, Emmons described rocks from Charlotte and Addison, Vermont. He concentrated particularly on the rocks at Snake Mountain, the southernmost member of the "Red Sandrock Range." North from Snake Mountain the range (today known as the "Overthrust Hills," so-called because they mark the line of the Champlain Thrust Fault) included Buck Mountain, Shellhouse Mountain, Mount Fuller, Mount Philo, and finally Pease Mountain and Jones Hill, before the Red Sandrock dove beneath the valley's soil to reappear less prominently at intermittent spots like Willard's Ledge, a quarry in Burlington.

Emmons's geologic education consisted of an essentially "Wernerian" view of the earth's history. Abraham Gottlob Werner, a late-eighteenth-century "geognocist" from Saxony, had summarized earlier ideas of a worldwide succession of rock strata deposited sequentially during the gradual subsidence of a primordial sea. Werner separated all of Earth's rocks into five classes—Primary, Transitional, Floetz, Alluvial, and Volcanic. This classification crossed the Atlantic with William Maclure, a Scottish immigrant who became a prosperous merchant in Philadelphia. Maclure published the first geological map of the United States in 1809, deviating only slightly from a Wernerian classification: On Maclure's map the Atlantic and Gulf Coastal Plains were colored yellow to signify "Alluvial" rock; New England and the Blue Ridge and Piedmont sections of the Appalachians were colored pink for "Primitive" rock; linear streaks of red up the Hudson, around the Narragansett Basin, and on the western margin of the Appalachians, marked "Transition" rocks; and from this western margin west to the Mississippi and north to the Great Lakes, Maclure colored his map blue, for the "Floetz or Secondary" rocks.

By the time Ebenezer Emmons began tramping the Red Sandrock Range, the classification of Werner and Maclure had altered little, becoming, in Emmons's report, Primary, Transition, Secondary, Tertiary. For the New York geologists, the classification fit admirably, since the Adirondack massif presented itself to them as an ancient island, around whose borders a succession of sediments was deposited. Receding from the ancient shores of that Primary island mass, they saw themselves as passing from older to newer "Transition" deposits. The base of these was a hard sandstone (exposed best in the town of Potsdam) that ringed the Adirondacks. This rock—the Potsdam sandstone—was believed by Emmons to be the oldest sedimentary rock on the globe, resting as it did on the Primary, unfossiliferous, crystalline rock of the Adirondacks.

Emmons and his colleagues on the New York Survey—Lardner Vanuxem, William Mather, and James Hall—had transformed American geology from the Wernerian stage of its development to a stage in which methods and standards were set for chronological delineation of the rocks. Their four reports replaced Werner's Transition series with a more detailed New York Transition System. Emmons called this "the most perfect system of rocks which has hitherto been described." From the base of the Primary Adirondacks, the unaltered, richly fossiliferous Paleozoic strata sloped gently away to the southwest in an unbroken series of gigantic steps, right up to the "Old Red Sandstone," the base of the Secondary. Emmons gave to the lowest group of the New York Transition System the name "Champlain Group," since all the rocks encompassed were found not far from the borders of the lake.

At the base of the Transition System was the Potsdam Sandstone, actually a "hard quartz rock" (i.e., quartzite), poor in fossils, and exhibiting in many places ripple marks that indicated that the sediments had formed in quiet, undisturbed waters. Above the Potsdam lay the Calciferous Sandrock, a sandy limestone, then the Chazy Limestone, marked by the appearance of three new fossils—*Maclurea, Trochus,* and *Columnaria*. Two other limestones—Birdseye and Trenton—followed, then the Utica Slate and Lorrain Shales, before ending with the Grey Sandstone. In contrast with these strata, the immense pile of the Taconic Mountains towering abruptly over the Hudson lowlands appeared to Emmons to be a distinct system of strata within the Transition. Emmons saw this Taconic System as the oldest, and first-formed system of sedimentary rocks, deposited on the Primary crystalline crust of the young earth. However, while in western New York the strata were simply positioned one atop the other, at the Taconic front in Massachusetts and Vermont the regularities of succession and structure evaporated, giving way to a confusion of highly folded rocks. Unlike other geologists, who had come to the conclusion that these rocks were simply metamorphosed masses of the Champlain Group, Emmons reasoned that the Taconic rocks, though *geographically between* the upper members of the Champlain group and the Primary rocks of the Berkshire and Green Mountain Ranges, lay *geologically below* the lowest member of the Champlain Group (the Potsdam Sandsone).

"Taconic," the name chosen by Emmons for the rock series, commemorated the Taughonic Indians of the hills east of the Hudson River Valley. This romantic designation of geologic systems after aboriginal groups was not limited to Emmons—"Siluria" commemorated a vanished European barbarian tribe, and "Cambrian" was after the name of the Roman province in Wales. American geologists variously used other aboriginal terms—Comanchean, Huronian, Algonkian—though none of them gained widespread acceptance. All of the names carried a sense of antiquity, of a primal past long vanished. Emmons envisaged the Taconic system as lying along both sides of the north–south running Taconic Mountains, passing through the New York counties of Dutchess, Columbia, Rensselaer and Washington, and through the whole length of western Vermont, as far north as Quebec.

Despite what the rocks told Emmons, he found few converts to his theory of a Taconic System. A roll call of the era's most prominent geologists told the tale: In America, James Hall, Henry D. and William B. Rogers, James Dwight Dana, J. P. Lesley, and others refused to acknowledge Emmons's Taconic (Lesley once dismissed it as a "fiction"). In Canada, Sir William Logan, director of the Geological Survey of Canada, and T. S. Hunt, his principal aide, unequivocally rejected any pre-Potsdam (Taconic) strata for the whole of North America. On two trips to the United States, Charles Lyell, the most prestigious figure in world geology, visited

John Bulkley Perry. Special Collections, University of Vermont.

the Taconic region and pronounced against Emmons's Taconic System. In 1850, at the annual meeting of the American Association for the Advancement of Science's Geology Section, W. B. Rogers pointed at Emmons, who sat in the audience, and said, "The Taconic System is dead, dead, dead!"

In 1861, two years before Emmons died, a Congregationalist minister of Swanton, Vermont, John Bulkley Perry, with his friend Dr. George M. Hall, discovered on Hall's farm in Georgia, Vermont, fossil trilobites that would help resurrect the Taconic controversy. Perry sent the specimens to Elkanah Billings, the paleontologist for the Geological Survey of Canada, who described the new fossils and concluded that they were further confirmation that Emmons's views were correct. In 1859, James Hall had described three species of another trilobite, *Olenellus,* from the same range of rocks as Perry's discovery, on the Parker farm, only a few miles from George Hall's. James Hall had pronounced these species to be of Hudson River (i.e., Upper Champlain Group) age. Billings, who had come quietly to his support of Emmons's Taconic because his superior, Sir William Logan, was allied with Hall against the Taconic, sent Hall's 1859 publication on the Georgia trilobites to Joachim Barrande. Barrande was a French royalist exile whose works on the rocks of the Bohemian Basin had established a firm geological column for the lower Paleozoic in Europe. When Barrande received Hall's publication, he recognized that the Georgia trilobites would challenge his "Primordial" life period (the base of the Paleozoic) if Hall's identification of them as of Hudson River age were to stand.

In May of 1860, Barrande wrote to Jules Marcou, a fellow European, to inquire how American geologists had determined the relations of these strata. Marcou was Swiss, and had come to America in 1853 under the auspices of Louis Agassiz. In response to Barrande's inquiry, Marcou traveled to Georgia to the Parker farm where the controversial fossils had been found. Mr. Parker sent Marcou to Perry, since he was the acknowledged local authority on the rocks. The meeting was fruitful for both men—Marcou was fortunate to find there, far from the learned halls of Cambridge, both an able geologist and a cultured individual in the person of the Reverend Mr. Perry.

Born in Richmond, Massachusetts, in 1825, Perry moved with his family to Burlington, Vermont, at the age of six. He had entered the University of Vermont in 1843, pursuing a course of study typical of the day—mathematics, Latin, history, philosphy, and natural sciences. Though interested in geology, as an undergraduate Perry was more strongly called by theology. After graduating in 1847 he spent three years in the South studying and teaching. In 1849, writing to a friend and former classmate, Perry wrote of the California gold rush more as a missionary than as a geologist:

> How many well-laid plans have been broken up by this California gold fever! . . . Excitement is good if it is of the right kind; it must be enthusiasm, an excitement regulated by reason, which is to bring any great project to maturity . . . the same mines will prove a curse to the nation. They are leading the public mind in the wrong direction. If reports are well founded, wealth will pour in upon the U.S., and people will become corrupted.

Young Perry was a bit of a prude, his perception of events like the gold rush colored by a deep devotion to God. While his peers rushed off to Sutter's Mill, Perry looked inward. "I have got a great work to do before I shall feel prepared to perform my part in life. Have I not been very slack in my duty, and thoughtless in respect to my own eternal welfare, and that of others? I have worked hard but not so hard as I should have done. O that I may live the rest of my life with more particular reference to the great end of my being!" That "great end," which was to "glorify God, and to prepare to enjoy Him for evermore in Heaven," seemed continually to guide Perry. He traveled toward it systematically, almost automatically. Method, system, adherence, doctrine. Each morning from five o'clock to quarter of six he studied Scripture and meditated; from quarter of six to six-thirty German; eight to ten Natural Science; ten to ten-thirty French; ten-thirty to eleven-thirty Greek; eleven-thirty to twelve Latin. In the afternoon, from two to three there was history, followed by an hour each of English poetry, German, philosophy, and at the close of day he returned to the Bible and his devotions: "My devotional exercises are those I would least neglect. They put me in a happy tone of mind for other inquiries. But it is not on grounds of expediency alone that I would devote a portion of time to his worship. I would hope that I am influenced . . . by love. I would hope that I do it for the purpose of honoring and glorifying Him, not in hope of a reward."

Perry's piety led him to enter Andover Theological Seminary in 1850, not as monk, but more as modern, meditative man. For Perry, the nature and character of God were learned from two books—Nature, and the Revealed Word. At Andover, he would read doctrine and grapple with the great questions of theology, but in Nature he would be confronted with even greater questions of existence. The earth for Perry was covered with the tracings of God's finger, and he saw it as his mission to come to know these *works* of God as well as his *word*. On his twenty-third birthday, he wrote in his diary: "If my life is spared, I trust I may some day be able to reconcile the sciences with each other, and especially with religion. I am beginning to look upon that as the great work of my life. It is more than has yet been fairly accomplished, so far as I know, and more than I can hope to do satisfactorily." This was *the* great task of the mid to late nineteenth century, the reconciliation of faith with knowledge. The bitterness of the Taconic controversy revealed just how crucial empirical knowledge had become, and simultaneously presaged the decline of an older system of knowledge about life's origins.

Not until Perry's thirtieth birthday, on December 12, 1855, did his life's work truly begin. On that day Perry was ordained and installed as pastor of the First Congregational Church in Swanton, Vermont. Perry soon made the acquaintance of Dr. G. M. Hall, a Swanton physician, and together they spent their free days investigating the rocks of the Swanton–Highgate area. When Marcou reached Perry, the Swanton minister had been studying the "Taconic" rocks of the area for five years.

For John Bulkley Perry, geological revelations in the field were not unlike God's revealed word. Stay fit in mind and body and spirit, and one's inquiry, whether spiritual or scientific, would be met with Revelation. The meetings of his mind and God's work took place all over northwestern Vermont, but some of the most poignant were within the Lewis Creek watershed. Of the Taconic rocks, it was the Red Sandstone that most interested Perry. He had grown up not far from Willard's Ledge, in Burlington, where the colorful quartzite was quarried, and it formed the foundations of many of the buildings that he had known since his youth. He had examined this "reddish or cream-colored ribbon" of rocks, as he called it, along much of its length—in Ferrisburg, Charlotte, Hinesburg, Monkton, Starksboro, Burlington, and Colchester—and he had even more intensively studied what he took to be its northern extension in Franklin County (actually the upper sandy members of the subadjacent Dunham dolomite). Of all of the places where the Red Sandrock was exposed, none was as dramatic as "The Oven" in Monkton. Located on the J. A. Palmer farm, just south of the Charlotte town line, and west of the Hinesburg line, this remarkable plication of rock had attracted the attention of geologists since the 1840s, when Charles B. Adams, the first state geologist of Vermont, had described the locality in his *Second Annual Report*. In 1854, Zadock Thompson came to the spot, followed by A. D. Hager, Charles H. Hitchcock, and, in 1867, Sir William Logan. All of these men were accompanied to the "Oven" by Henry Miles, whose own farm lay less than a mile away.

The little north–south ridge that holds the "Oven" is wooded now, and is most easily approached from its more gentle eastern side, which is a dip-slope. (That is, the slope of the land approximates the same angle as the dip—the number of degrees away from the horizontal—of the underlying bedrock.) There is a road leading to the top of the hill, where some years ago a young man had built his dream house on the ledge, before it was struck by lightning and burned to the ground. One can stand among the blackened timbers and broken mortar, in a little grove of staghorn sumac, and look north along a narrow bald lane of grasses and blueberry, flanked by scrubby red oaks and shadbush. Directly south lies the cedar swamp that Henry Miles used to mine occasionally for peat, and at the back end of this, a pair of unnamed peaks cradle the summit of Mount Florona. Two

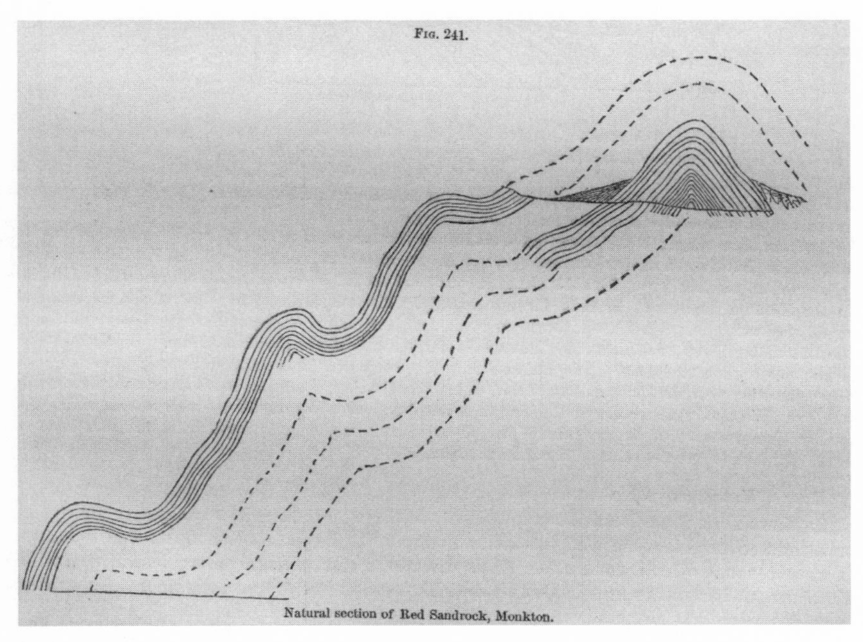

"The Oven." From Edward Hitchcock et al., *Report on the Geology of Vermont—Descriptive, Theoretical, Economical, and Scenographical,* 2 vols. (Claremont, N.H.: Claremont Manufacturing Co., 1861).

steps toward the south, the ledge drops off about ten feet—this is the "Oven," now a less remembered landmark in Monkton than the house ruins that sit atop it. It is immediately obvious how the spot got its name. Actually the crest of a huge fold, some of the lower strata at this crest have been eroded away, leaving a perfectly arched cavern that resembles an old-fashioned brick oven. Porcupines have succeeded in making a home here, filling the cavern with mounds of their droppings.

All of the facies of the Monkton Formation/Red Sandrock are displayed at the Oven. There are two-foot-thick beds of yellowish-orange dolostone cut by small calcite veinlets, and thin red shale partings separate the beds. Above this there is more shale, with the layers thicker than below, changing to a fine sandstone. Above it all is the red quartzite that gives the formation its name. That the beds were so clearly visible was not so remarkable; that it displayed the fold so perfectly was. The "Oven" was a window onto the forces that disturbed all these rocks that once lay as horizontal beds.

What were these forces? Though Perry in his writings talked often of "elevating forces," the "lifting up" of strata, and so forth, he never speculated on the mechanics or causes of such forces. If he had subscribed to the

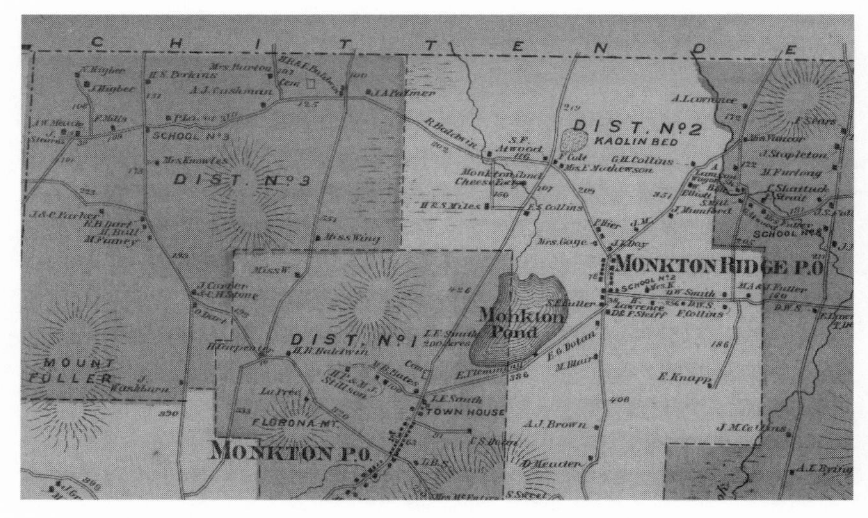

From F. W. Beers, *Atlas of Addison County, Vermont* (New York: F. W. Beers, A. D. Ellis, and G. G. Soule, 1871).

view of the Rogers brothers, he would have envisioned huge volcanic eruptions shaking the deep sedimentary cover as if it were a blanket being shaken at one edge, the folds in the blanket preserved as the underlying molten material solidified. Or he may have thought more along the lines of James Dwight Dana, who believed that the various folds of the Appalachian system were caused by a cooling Earth that shrunk as it cooled. Like a piece of fruit, the surface of the sphere wrinkled as it shrunk, and the "Oven" ridge was simply a limb of one such wrinkle.

The explanation of cause may have been lacking, relying on weak models of shaken blankets and shriveled fruit, but the depiction of effect sufficed. Perry could say what he *saw,* even if he could not say what caused it. In order to fit the rocks into their theories (so that the Red Sandrock was equivalent with the Medina sandstone), many geologists supposed the "Taconic" rocks to have been completely overturned during their uplift, the bottom strata coming to lie at the top and vice versa. But Perry had gazed at the rocks all over Charlotte and Starksboro and Monkton, and a simple observation told him otherwise. At a number of outcrops, Perry found that the Red Sandrock contained impressions of raindrops, *in their normal position.* Strange scenarios had been postulated in nineteenth-century geology, but none said that rain fell from below!

The most convincing evidence that Perry found by looking through these outcrop "windows" was the fossils he found in them. In 1861, Elkanah Billings described Perry's fossil discoveries. There was *Paleophycus incipiens,* its Latin name tenuously suggesting that it was the "beginning to become, ancient seaweed." Its thin, unbranched, curved stems spoke of a simple plant, one at the beginning of that kingdom's history. In similar rocks Perry found several brachiopods: *Obolella cingulata, Orthisina festinata,* and *Camarella antiquata.* Most importantly, Perry found two trilobites—*Conocephalites Adamsi* and *Conocephalites teucer*—the crustacean creatures that so often came to serve as index fossils for periods of geologic time. These were all "Primordial" life forms, plants and animals that sprang from a dark evolutionary past into the fossil record at the base of the Paleozoic. To Perry, they were not "Medina" in age, for they were without exception extinct by that time.

There was another fossil, not discovered by Perry but by Henry Miles, which Perry considered to be the very oldest life form. It was a little clam, a *Lingula,* which Miles found on Truman Hill's farm in Starksboro, near Rockville. The fossil was from a boulder, not a bedrock outcrop, so its precise stratigraphical origin was unknown. Perry and Miles scoured the area for additional specimens, both in the boulders laying out in Hill's pasture as well as the bare grey ledges of quartzite, but found none. The little *Lingula* fascinated Perry; he reasoned that this quartzite was at the very base of Paleozoic time, and yet frozen in it was an advanced life form, a brachiopod, appearing "at the very dawn of life on earth." This was Perry's true quest, not to construct the correct chronological column for the rocks, but to get at the nature of that dawn. The Taconic was the "first great system of Life" to Perry, and through the window of the Taconic rocks he hoped to peer at his Maker.

During the same time that Perry had been promulgating contrary scientific ideas regarding the Taconic rocks, he had been preaching religious doctrine that struck the Calvinist Congregational elders as heretical. By the end of 1865, the pastoral relation between Perry and the Swanton congregation had been severed, and on January 7, 1866, Perry preached his last sermon there. He spent some time in Hinesburg with his friend Clark Ferrin, and at the homes of his wife's families, the Leavenworths and Willsons. Perry had married Lucretia Leavenworth Willson in March of 1856, but a year later Lucretia died in childbirth. Their son, Francis Willson Perry, was raised by Lucretia's mother in a house just across the Lewis Creek–La Platte

River divide. Like so many large nineteenth-century farmhouses of the Champlain Valley, it stands up above the blue clay soils on a bedrock knoll. The little patches of second-growth woods that usually flank these farmhouses harbor bedrock outcrops, the windows that always attracted Perry's eye.

Perry's journals note his visits to the outcrops by the Willsons' house, but they contain not one word of the wife he lost, or the son from whom he was separated. Though he was sensitive and spiritual, his grief and joys were always private, kept even from the paper where the rest of his thoughts always appeared. One of the few places where any emotion surfaces at all in John Perry's diaries is in the early spring after his dismissal from Swanton. On Thursday, April 5, 1866, there is an entry in a different ink and a different hand from the rest: "Mild, beautiful spring day. Walked with M. P. on the lake shore. Gathered mosses, observed the rocks, etc. All together found it very delightful. M. P." This entry is followed by John's detached tone: "The last observation was made by sister Maggie—on her own account alone—tho it was fully expressive of the mind of her brother John, who accompanied her." A week of similar entries follows, Maggie writing poetry and speaking of their happiness, John noting the appearance of the lake ice, the date of its breaking up, and the direction of the wind.

By June, Perry had been appointed delegate to the United States Christian Commission, and later that year served as chaplain of the 20th Vermont Regiment. In his diaries from this period, interspersed between notes on the suffering of the soldiers and his own day-to-day doings, is a wealth of geological observations. At the taking of Petersburg and the surrender of General Lee at Appamattox, his comments on the rocks take precedent. His reference to "Taconic" rocks in some places shows that he carried south the stratigraphic vision that he had developed observing the rocks of the Champlain Valley. After a brief period in the winter of 1867 ministering and teaching in Wilmington, Vermont, Perry moved to Cambridge, Massachusetts. Whether it was the desire to be closer to the centers of geological or theological inquiry is difficult to know. Perhaps it was partly due to the fact that Perry had in May of 1867 remarried. In June of 1867 the Reverend Mr. Perry and his wife moved to Cambridge, and along with the weekly meetings of the "Society for Religious Inquiry" that the couple attended, Perry attended meetings of the Boston Society of Natural History. On December 18, 1867, Perry addressed the group with a lecture entitled "Queries on the Red Sandstone of Vermont." In it he reviewed the history of views about the Red Sandstone, and gave his own views on the subject.

Perry was unknown to the majority of naturalists convened at that December meeting, including Louis Agassiz, the Swiss naturalist who in 1859 had founded the Museum of Comparative Zoology at Harvard University. At the close of Perry's address, Agassiz rose and spoke of his interest in the Taconic question and of Perry's able presentation. Agassiz recognized the originality of Perry's views and the detail of his observations, saying that he knew now who furnished other geologists with the materials that they used in their discussions of the topic. At the close of the proceedings, Agassiz invited Perry to the museum and in short order offered Perry a position as paleontologist.

Since 1859 when Darwin's *On the Origin of Species* was published, Agassiz had become more and more ostracized by America's leading scientists, particularly his coeditors at the *American Journal of Science,* James Dwight Dana and Asa Gray. Even to his young assistants at the museum, Agassiz had become "the Great Annihilator" due to his opposition to the theory of evolution, and museum scientists E. S. Morse, A. E. Verrill, F. W. Putnam, and others had by 1864 left for other institutions. Agassiz had replaced his assistants with Europeans like Jules Marcou, men who might more easily be tethered under Agassiz's direction. It was probably as much due to Agassiz's recognition of Perry's humility and theological background (and hence presumably less radical views on the question of evolution) as to his recognition of Perry's geological abilities that prompted Agassiz's offer to Perry of the paleontologist position.

Whatever Agassiz's motivation, Perry was afforded an opportunity to continue his investigations on the "first great System of Life." He came to the museum at a time of major reorganization of its paleontological collections, and set to work on classifying and arranging the lower Paleozoic fossils. Going to work each day at the greatest natural history museum in the country put Perry in a strange position; he was poised between ideological forces of tremendous strength. Though Agassiz's public attacks on evolution had ceased, he had spent years as the preeminent anti-Darwinian voice among the American scientific elite. Seven years before Perry spoke to the Boston Society of Natural History, Agassiz had strenuously advanced the classic defense of special creationism—that God had individually and uniquely created all creatures, living or extinct. Agassiz had been a bear of a man who commanded the attention of all who met him, and as interpreter and popularizer of science he was the scientist to whom America looked for guidance in the first tumultuous years following the publication of *Origin of Species*. Now almost no one looked to him; Perry came to know a great man in his twilight years, his strength diminished, his fire cooling. Two months after he wrote out a detailed list of his objectives for Perry's work at the museum, Agassiz was struck by a heart attack. In September, after a summer of rest, Agassiz was to give the address at a meeting of the Boston Society of Natural History celebrating the centennial anniversary of the birth of Alexander von Humboldt, explorer, geographer, and naturalist. Agassiz had been von Humboldt's student, and was a link between von Humboldt's pioneering and the beginning of modern natural science. Later that day Agassiz suffered a massive cerebral hemorrhage, leaving him completely paralyzed and unable to speak. Though in time he recovered, Agassiz's day was done. His faith in the Great Chain of Being was unshaken, but the rest of the world was undeniably exchanging that faith for one that included evolution by means of natural selection. Even Perry, whom he had chosen to arrange the first precious links of the "Great Chain" as they were represented at the museum, had a more Darwinian than Agassizian view of the universe.

Along with his curatorial duties, Perry did some teaching, both at Harvard and at his alma mater, the University of Vermont. It gave him a chance to organize the body of knowledge he had acquired in the field and in the classroom. In preparing his lectures he showed his propensity for outline making—every subject was broken down and numbered in relation to its companion topics. As in his writings, he elaborated every point, allowing no student to take anything for granted. Such elaboration may have been regarded as his chief fault—he often explained too much. Agassiz, in editing a paper Perry was writing for the *American Journal of Science,* told Perry "everywhere to be more direct and less apologizing" and to "condense, condense, condense and then erase every unnecessary word and phrase."

A survey of the lecture note outlines makes a neat condensation of Perry's interests. For an introductory course in geology, Perry followed this outline:

I. Introduction of the Globe (or Primogeniture of the Earth)
II. Encrustation of the Globe
III. Composition of the Globe
IV. Disposition of the Globe

Beginning with speculation on the nebular hypothesis and other earth-formation theories, Perry took his students through discussions of the internal heat of the earth and the mechanisms of maintaining it; he gave an introduction to rocks and minerals, to the law of superposition of strata, and to the principal geological formations; he explained the elevation of rock masses and their subsequent denudation. An equally neat package was made of "The Successive Systems of Life." In Perry's tabular view of the rocks of the globe, there were three types of "Azoic" rocks (though Perry noted that one should "not take it to mean destitute of organic existence, but instead that no fossils have as yet been found")—molten, massive, and foliated—and three principal ages of "Zoic" rocks—Paleozoic, Mesozoic, and Cainozoic. Keeping with his belief in a "Taconic" system of rocks, the Paleozoic was preceded by a dim "Protozoic," comprised wholly of the Taconic era.

Louis Agassiz gave his own tabular view of how Perry's writings should be arranged:

1. Theological Geology
2. Tertiaries
3. The Lake Champlain Series
4. Massachusetts Geology
5. Glacial Phenomena
6. Paleozoic Corals
7. Foliated Rocks
8. Change of Level of Continents

That the crowning topic chosen by Agassiz was "theological geology" was no special bias of Agassiz's—the theme of God's word as related to God's work was always apparent in Perry's writing. In his introductory course, he closed with a lecture on the "Relation of Geology to Scripture." Readier than his mentor Agassiz to accept Darwin, still the Great Chain of Being held a powerful hold on his mind. After surveying the whole of the animal kingdom in the most rigorous scientific manner of his age, he closed:

> And finally, we may bow with reverent humility and childlike wonder, as we survey these vast displays of intelligence and power. All the comprehensive systems of life which have passed in review before us, are the work of Creative Wisdom. Their several places have been exhibited upon the face of the globe from age to age during the long and unmeasured past. Each in due order was called into existence, then entered upon its appointed work. With the exception of the last, they have all finished their course and disappeared.

> Their career is ended. Having acted their part, they left the stage of existence. Their destined race being run, they are now known by only a few remains. As these, from time to time, are exhumed from the crust of the earth, they tell a weird and mystic tale, leading us to look upon the envelope of the whole globe as the vast repository of many preceding creations. They teach us to regard it as the grand mausoleum of manifold forms of life long ago extinct, and thus as the great burial place of the countless dead. The sediments which were once their graves . . . are turned to stone. It has come to pass that the old inhabitants of the earth now live entombed in marble; their frail forms have become their lasting monument, preserving their life and death and epitaph in words of light, both giving back their long history, and handing down to distant times their strange memorial.

When Perry said that each system of life was "called into existence," it seemed that he stood on the side of special creation, yet it was not that simple. Other scientists tended to put forth their beliefs in a way that made them "for" or "against" the theory of evolution by natural selection. Among these men of science, Christian commitment was not the exception, but the rule, and both in North America and in Europe, scientists led the opposition against Darwin's theory. The response of the religious community was more positive—many leading theologians were very receptive to the new view of life. Congregationalists showed perhaps the greatest affinity for evolution, but few were qualified to rule on the *scientific* as well as the *theological* merit of "Darwinism." Perry certainly was. By a circuitous route he had come finally to the place where his youthful idealism had pointed—a position of reconciling science and religion. In Perry's lectures on evolution, he seems at home in a land between the two conflicting outlooks: "Some take the ground that God acted before all time. Others that he acts through all time. There is trouble in both positions. The error is in assuming one and excluding the other. Properly we must hold to both alike and exactly." In determining a hypothesis to show how species originate, Perry believed there to be two principal facts: "1. Each species contains much in common with preceding ones. . . . 2. Each species contains a principal superadded to what precedes, and distinctive of the species itself." Perry thought it natural to infer that there was a material connection between "higher" and "lower" creatures. In each organism there was a "new principle, a creative fiat, set in operation. One may say it was potentially present in the lower species, not as distinctive of it, but as the residuary creative force which would in due time become operative according to the divine purpose. In this view, it is a new creation." While others debated the fossil record, compared bones, and quoted Scripture, Perry intuitively invented a black box—his "superadded principle."

Throughout the twentieth century, from Bergson's *élan vital* to W. M. Wheeler's "emergent evolution" to Lovelock and Margulis's Gaia hypothesis, science has coined new superadded principles to counterbalance its thrust toward reductionism. But there is no place left for Perry's God in even the most theologically sensitive of contemporary evolutionary theories. No geology students who visit the Oven today and pluck brachiopods and trilobites from the red layers with their rock hammer see the Great Chain of Being emerging in the Cambrian muds. When nongeologists canoe down Lewis Creek and, skirting outcrops of Monkton quartzite, are occasionally set to reverie about earth's origins, rarely do their thoughts resonate with the need to reconcile the rocks with Scripture. The rocks and their fossils have largely *become* Scripture, so that John McPhee's phrase "deep time" sounds as numinous to us moderns as the word "God" did to Perry and his contemporaries.

When it came to the most thorny of issues, humanity's place in nature, Perry also reached for a "superadded principle." In his notes for a lecture on "The Origin and Antiquity of Man and the Unity of Race," he recognized that there are essentially two views: evolution from anthropoid primates versus direct creation by God. His reasoning derived from only two facts: "1. Man has an animal nature. This he has as much as any brute of the field; 2. Man has a spiritual constitution. This he has as a superadded principle." Perry said that both are true, that man is both animal and divine. He recognized that both camps would object to such a view, that

> this allows that man is a divine creation. Most surely it does. The denier of this point must surely make his denial on the fact that the animal is predominant in him, the spirit largely suppressed. Others may object that it allows a lineal descent from the ape. . . . On this point it neither affirms or denies, for we do not and can not know. In affirming the creative addition to the animal of that which is distinctive of man as man, it holds that in some way the higher was planted in the lower, and came to be characteristic. Take even the Mosaic account: man made of the dust of the earth: dust is the earthly element . . . inorganic, next it appears in the vegetable, once more it

> occurs in the animal form. Now what more natural than the creative principle wrought in the highest (until then) animal form, that the divine breath wrought . . . this animal dust to its own higher ends, and as the result of this creative work, stood forth man, in "the human form divine."

So, by words alone, Perry seemed to bring reconciliation to the irreconcilable. His technique was more metaphysical than scientific, inductive rather than deductive. It was not because he was vague about the new science—he had studied Darwin's work thoroughly. His notes indicate a close familiarity not only with *Origin of Species,* but with the many scholarly and popular articles that came in the wake of its publication (most notably Agassiz's Lowell lectures and Asa Gray's series in the *Atlantic Monthly*). He believed Darwin's theory merited careful and candid weighing, and was impatient with the summary judgements passed by many of his peers "upon a work which cost twenty of the best years of his life." Perry was able to accept the whole of Darwin's theory. It was what Darwin omitted that disturbed him: "The grand error of Mr. Darwin is not that there is slow gradation . . . but that new species . . . come merely from this; that he does not insist on a superadded principle, no matter how slow its manifestation."

Perry's reconciliation of science and Scripture was much easier for his Cambridge colleagues and congregation to accept than his heretical doctrines against eternal damnation had been ten years earlier in Swanton. There was a wide forum for such views as his. While he continued publishing his research in geology in journals like the *American Journal of Science,* he voiced his theological thought in *Bibliotheca Sacra* and the *Congregational Quarterly*. In April 1870, the latter magazine ran a piece by Perry entitled "A Discussion of Sundry Objections to Geology." The article refuted point by point the various criticisms made of geological science—that geology is made up of extravagancies, inconsistencies, and baseless theories; that geological deductions are premature; that the Noachian Deluge was sufficient to account for all the main points of geology; that the fossils relied on as evidence were created in the rocks as they were found. This last claim brought out Perry's most vigorous prose, for Perry was first and foremost in science a paleontologist. Fossils were not mere curiosities to him. They were the stuff of his science, familiar and particular. He knew fossils as objects variously in greater and lesser states of petrifaction. He knew fossils as evidencing the adaptations that their possessors would have required in their specific environments. He knew fossils as objects whose portions were often worn away by use or damaged by mishap. He knew this as true of not just a few exceptional instances, but as applied to the whole creation, vertebrate and invertebrate, freshwater and marine, ancient and recent.

Those who felt that the living forms of their own time were real products of organic life, while those of all preceding eras were just semblances, were ignoring the obvious. Perry pointed out their inconsistency by way of analogy:

> Let a boy pick up, amid old rubbish, a knife bearing the impress of 1775, a knife the blade of which, having been made of good stuff, had been used until it was ground to a stub. Now convince him, if it be possible, that the old blade which he holds in his hand was not made for cutting; or to carry out the analogy still more closely, that all the contrivances ordinarily called knives, which were manufactured previous to the year 1800, while very much like those of today, were never intended for any of the practical purposes for which such implements are now employed.
>
> This, as we all know, is plainly out of the question. So take a skillful paleontologist, who finds in a fossil condition . . . a worn tooth or a broken tusk of an animal belonging either to an extinct or to an existing species . . . and convince him . . . that the fragment which he holds in his hand, while it exactly resembles a tooth, a tusk, or some other fossilized organic part, was never used as such, and in no wise ever formed a constituent portion of a living creature. He sees the thing somewhat as it is in itself; not merely as an isolated outward object, but in the light of many closely connected principles involving manifold necessary relations; and his convictions consequently rest on an immovable basis. The conclusion, of course, is evident and inevitable. So is it everywhere.

Perry went on to the objections raised against geology because of its supposed contradiction of the Mosaic account of Creation. He examined the meaning of the word "day" and fit it into the geologist's deep view of time. He then interpreted each of the successive days of creation geologically: the first day, the combined creation of the heavens and of the earth; the second day, the creation of the solar system; the third, the formation of dry land and the introduction of plants; the fourth, the sun appearing through the thick vaporous atmosphere of the young earth; the fifth, the animal kingdom's development; the sixth, the appearance of mammals and Man, the crowning work of creation. In a closing footnote, Perry suggested that a distinct department for examining the relations between the sciences and the Bible be formed in the nation's theological seminaries. Within a few months he received an invitation from Oberlin College to found such a department there, and in July of 1871 was appointed Professor of Geology and Natural History and Lecturer on the Relation of Science and Religion. Perry insisted that geological field work be included in the new course of study: "Without practical exercise of one's powers on the rocks themselves, no one can become a skillful geologist, I might also say, even an intelligent investigator of the structure of the earth." He also stated the necessity for his own involvement in original research and field work, undoubtedly still related to that "first great System of Life" that had always intrigued him. Indeed, he had most recently published a study of the "Eozoon" limestones of eastern Massachusetts, believed by some to contain the oldest known fossils.

After five months of teaching at Oberlin, Perry was preparing to return to Cambridge to continue his work at the Museum of Comparative Zoology. He took a field trip with a few friends to a cave near Dubuque, Iowa, to gather specimens, planning only to spend a few hours there. The group emerged from the cave ten hours later, and though Perry was exhausted, he preached twice that day in Dubuque. He continued east, preaching the next Sunday the theme "God in Creation" before a congregation at Humboldt College in Springvale, Iowa. The next Sunday he was back home in Cambridge, but home only to die, for he had contracted typhoid fever. As the fever grew more violent, he remained preacher and scientist; he would examine specimens, then lay them aside while he prayed. In his delirium he persuaded some imaginary doubter to come to Christ, urging strongly the great truths of the Bible and of the Savior.

The day before he died, Perry fell asleep, then suddenly awoke, his face bright with joy. "Enchantingly! Entrancingly!" he cried. His wife asked him what was so beautiful. "Oh! All about us!" he replied. At that moment John Bulkley Perry seemed to have truly found a window from the physical world to the world of spirit and, looking through it, was awed by the scene. His persistent inquiry into the physical nature of the earth seemed to have fully prepared Perry for the moment of leaving it behind for an unseen landscape.

CHAPTER 9

Sebamook

ED BY DOZENS of rivulets and larger tributary creeks, each reaching up into their own unique pasts, Lewis Creek has but two bodies of standing water that feed it—Monkton Pond and Bristol Pond. Hopeful boosters of the area have tried to glorify Monkton Pond by calling it "Cedar Lake," and though cedars aplenty it once had, the ovate pool nestled at the western foot of Monkton Ridge is no lake. Hardly more than five feet deep in most places, Monkton Pond is just that—a pond. The word "lake" conjurs images of the sublime, while "pond" presents something only picturesque at best. Monkton Pond never inspired poetry, or attracted intellectuals to its shores for quiet contemplation. It holds no deep mysteries, and its relation to man has always been of a very prosaic sort.

Henry Miles, the Quaker miller-farmer who was John Bulkley Perry's guide to the redrock window called the "Oven," lived about a ten-minute walk away from Monkton Pond. Sometime after settling there in 1843, he became interested in the shell marl deposit that blanketed the bottom of the pond. The marl deposit is essentially calcium carbonate produced from the partial decay and crumbling of innumerable fresh water shells. Intermixed with the lime are small amounts of sand and clay. Ever the scientific farmer, Henry Miles set out to investigate the benefits of Monkton Pond's shell marl as a fertilizer. With two of his neighbors, Miles drew several sleigh loads of the marl and spread it on nearby meadows, hoping to compare the growth of the grass there with adjacent untreated meadows. The results showed that their generosity did "neither good nor harm," and Miles concluded that this was because there was already sufficient lime in the underlying soil.

Though his own fields did not need it—Miles's and the surrounding farms were underlain by dolomite, which gives rise to sweet soil—Miles was well aware that there were many Vermont farmers whose lands were not blessed with sweet, limy soil. His concern was that these farmers, and those who had already reduced their productive soils to barrenness by excessive cropping, have access to an inexpensive source of lime. Shell marl, he believed, could provide lime for fertilizer at one-fourth the cost of other fertilizers. Speaking to a group of farmers assembled in Hinesburg for a meeting of the State Board of Agriculture in 1875, Miles assured them that shell marl was no "Emma mine," promising much and yielding little, but a natural treasure that would cost nothing but the labor of digging and drawing. Edward Hitchcock, state geologist, concurred with Miles as to the value of lime, saying it was "a treasure which Providence has hidden in the earth, and provided for its elimination at the right time and quantity, and is of far more value, in my estimation, than all the subterranean wealth of the State."

Providence and Science—for Miles and his fellow farmers there were now *two* gods. Providence provided; Science improved upon it. Science gave the farmer a touch of control over Providence, transforming him from passive steward of God's green things to an active agent in nature. "The intelligent farmer knows that there is no *real* conflict between science and labor," said Miles to those old-school farmers in his audience who were skeptical about science, suspicious of agricultural colleges founded on the premise that there was more to good farming than early rising and hard work.

By the account of Charles B. Adams, the first state geologist of Vermont, there were about 300,000 cords of shell marl in Monkton Pond. Bent on extraction of this resource, Adams still gave pause to consider the fecundity necessary to produce such a deposit. The shells for the most part being less than a quarter of an inch in diameter, Adams figured there were more than six billion shells in the Monkton Pond marl. He also calculated that the ten-foot-deep shell sediment took at least twenty thousand years to accumulate. Since Adams was first and foremost a conchologist, his main interest was in just what types of molluscs were found in the marl. He found eleven species, all of them still living, chiefly of the genera *Paludina, Lymnaea, Physa, Planorbis, Cyclas,* and *Pupa*. Speaking of one particularly diminutive species of this last genus (*Pupa milium,* which is only three-hundredths of an inch wide), Adams once remarked that "none but a naturalist would find it."

Monkton Pond and the rest of the watershed, and the great world beyond, held much that none but a naturalist would find. The pond acted as a sort of sink for history, its marl bed encasing the geologic past, and the peaty muck above the marl holding narratives of the plant life that once thrived along its shore. Miles's friend John Perry sought to read the pond's history, and in doing so wrote to Miles for help. Perry was writing an article on the Pleistocene history of the area, and wanted to procure specimens of the molluscs in the pond, but was in Cambridge, Massachusetts. In December of 1868 Miles wrote to Perry to tell him the results of his searches for shells on the shore of the pond:

> I have searched the margin of the lake N. and S. Winds had thrown up some shells from parts distant from the shore. If I remember correctly we found only 2 varieties when I had the pleasure of thy company. I think I have added 4 more, if the pearly specimens that have the plan of the *Planorbis* be really of another class. I have enclosed some of the muck thinking it possible it may contain minute shells that I have overlooked. Some of the young ladies of the "Ridge" [Monkton Ridge] have from time to time engaged fishermen, when on the lake to procure shells for fancy work.

A bit of "fancy work" is one of a handful of specimens that has come down through the years as material evidence of Henry Miles's eclectic interests in natural history. His great-great-granddaughter, Mary Lee Rose, lives down the road from the Miles farmstead, and though her naturalist ancestor's collection was sold at an auction more than twenty-five years ago, she still has a small cardboard box of treasures collected by the Quaker farmer. What remains are a set of unlabeled objects that seem like random curiosities collected on a series of vacation trips—a piece of coral, some freshwater shells, a halite and a quartz crystal, a couple of "claystones" (concretions), a lump of iron slag, and a bit of chrysotile asbestos. There is another object of a different sort in the box, a green embroidered ball and bow attached to the shell of a freshwater clam. This is a piece of "fancy work," a clam collected by one of the fishermen wedded to the handiwork of some farmer's wife.

Henry Miles's dream that Monkton Pond's shell marl would make fertile the land of hundreds of Green Mountain farms never came to be. Too many marginal hill farms were already being abandoned for new lives on Midwestern prairie soil, and the chalky deposit never came to dry out in the summer sun to help grow clover and corn and timothy. But a few shells did make their way onto foreign soil via fishermen and fancy work. Some are still poised, not yet returned to earth, hanging in dark closets on family heirlooms.

As Monkton Pond came to be the burial place of innumerable shelled creatures, so did the land near its shores become a burial place for the farmers, preachers, millers, and others that lived out their lives within sight of the pond. Up on the Ridge Road, next to the Town Hall and overlooking the pond, there is a cemetery that holds the Bulls, and the Catchapaws, an Abenaki family who farmed a piece of the watershed due north. Their place is part of the Baldwin farm now, Lyle Baldwin's favorite "piece." He remembers as a teenager when he and his brother used to wander down to George Catchapaw's for a few sips of George's hard cider, and how, though everyone liked to think of George as a white man, there was something different, something definitely not white about George. In the woods, trapping fox or hunting deer, George was more at home than his white neighbors, moving with the quick and easy assurance of a woods creature. His death certificate lists "white" as his race, but the undertaker knew better.

George's daughter Maude and her husband, Frederick Lawrence, are buried here. Maude died young and left Frederick a widower for 37 years. Their stones and all the others in the Ridge Cemetery are presided over by Ammi Fuller's monument, rising 15 feet from the center of the burial ground. Next to Fuller is the Mumford plot, one of whose inhabitants sports this cheery epitaph on his headstone:

> Death is a debt that's ever due
> Which we have paid and so must you.

The grim reaper is stayed on any June day there, when wild strawberries run rampant over the graves, and chestnut-sided warblers sing from the hedgerow, and chimney swifts twitter their insane chuckle above it all.

Over on the slope above the main feeder stream to the pond is another cemetery, the headstones all facing east toward the pond. All the cemeteries in the watershed have east as their cardinal direction. Only stones from the last half century are oriented any other way. Before World War I, all of Lewis Creek's dead who were hopeful of a hereafter faced the rising sun. In the morning, their limestone headstones shine as brilliantly as the open, quickwater stretches of the creek. Seen from one of the Cheshire quartzite bare spots of Hogback Mountain, the tiny stones of the Smith Cemetery in East Monkton look like cubeiform bones strewn in the grass. From the open ledge on Lincoln Hill in Hinesburg, the Hollow Brook Cemetery's stones are even smaller, yet when they catch the morning sun they sparkle significantly out of a view that is perhaps the most dramatic in the entire watershed.

Northwest from the pond about a mile and a half, there is the most forgotten of Monkton's cemeteries—the Old Quaker burial ground. It stands on Ed Rotax's farm, surrounded by cornfields and marked at a distance by a couple of big white pines rooted amongst the graves. If you stop to ask Mr. Rotax for permission to cross his fields to the cemetery, he'll consent gladly, happy that someone is paying a visit to the place. The town does not maintain the cemetery, he says, and though he has tried to keep it from reverting to forest, he can't keep up. He'll tell you about a grist mill that was over in a nearby field, how the millstones are still there, how magnificent the woodworking in the building was when it stood, how the flume was sealed with blue clay. He'll tell you about the saw mill on Lewis Creek that sawed the white oak beams in his barn, 12 inches by 18 iches and 45 feet long, and lament the fact that these days most mills cannot saw anything over 12 feet. He talks of the antique cars, including his favorite, a 1915 Dodge, that lies rusted and broken behind the barn. If you ask about who's buried in the cemetery, he'll tell you how Vermont's first representative to Congress is buried here, and how the people of Monkton had to take up a collection to bring his body back from Washington after he died. He'll talk about the night he looked up to the ridge just east, to see a young "homesteader's" house burst into flame after the propane tank was struck by lightning. Its ruins are those perched atop the Oven; he has seen the charred remains of the house but has never seen the arched quartzite nor heard of John Bulkley Perry or Henry Miles.

Henry Miles is buried out there under the great pines, his headstone toppled over into the tall grass. It faces east toward the quartzite cliff summit of the Oven. Nearby lies a fellow Quaker whose visions far exceeded the use of marl beds, or anything else in Henry Miles's vernacular universe. His headstone surrounded by woodchuck holes, thickets of pasture rose, and last year's dead mullein stalks, Quaker "prophet" Joseph Hoag lies less than a hundred yards from the sunny spot where in 1803 he had a remarkable vision of religious societies dividing, including the Society of Friends; the Civil War, the slaves emancipated; and the emergence of a new monarchial power in America. "This Vision is yet for many days," a voice had said to Joseph Hoag. The Hicksite–Orthodox Quaker sectarian split in 1828 was the first fulfillment of Hoag's vision; he never lived to see the deeper division between North and South grow into war, though in

1846 when he died, one needn't have been a prophet to know civil war was inevitable.

One could follow the creek's tributaries out to all the watershed's cemeteries. Up the road from the Pringle farm is the Morningside Cemetery, holding Cyrus and Frank Estey and the families of Prindle and Baptist Four Corners. In the Methodist Cemetery at North Ferrisburg is buried Rowland Robinson Preston (1888–1952), named by his father Sedgwick Preston for Sedgwick's fishing companion. The cemetery is flanked on the west by a row of Scotch pine. A hundreds yards away, the intersection of the Hollow Road with Route 7 is marked by two solitary gray pines, transplanted outposts of the gray pine stand found by Robinson in 1876. There's the Collins Cemetery, where Hiram and Heman and Peter Bull are buried, and the Quaker and Old Quaker Cemeteries, where Hicksite and Orthodox are sometimes buried side by side.

In Hinesburg, two cemeteries lie within the Lewis Creek watershed. The Baldwin Road Cemetery, whose prominent names include Davenport, Partch, Burritt, Hayes, Elliot, Palmer, Bostwick, Page, and Ferris, is best visited in late May, when the moss pink turns it into a sheet of flame. The little plant has long lost its eighteenth-century signification of "a token of all the heart can give, of love in its fountain deep." The Creek's other cemetery in Hinesburg is the one that looks east up the Hollow Brook valley, toward the tip of Camel's Hump. The graves there were always easily dug, into the loose gravel left when the Huntington River drained into glacial Lake Vermont instead of running north to the Winooski. In that time Lewis Creek was a real river, its icy waters carrying rock flour from the ice-ground Front Range Hills, its channel never reduced to the shallow stream that now runs any August day. That raging river is preserved only in the deep notch in the Front Range escarpment, and in the deltaic deposits that make such good road fill and easy digging for graves. The huge gravel quarry there exposes the structure of the sediments that spilled into Lake Vermont, showing them to dip gently west. Imbedded in the strata are the Pleistocene's own dead—carbon traces of fossil ostrich fern, petrified logs, and fish bones.

Down here on the flat, fertile valley soil, the headstones are much finer than those of the Front Range cemeteries of Hillsboro and Little Ireland. Here there are Sweets and Algers and Calkinses and Youngs and Kinsleys. One tall obelisk marks the grave of James Albert Kinsley, who died in California on February 7, 1853, at age twenty-nine, surely one of Vermont's contribution to the Gold Rush. On any October day monarch butterflies will light on the stone's pyramidal top, resting momentarily before continuing their southward migration. The monarch's is an ambling passage, much like that of humans—not purposeful nor directed, but dispersed, ambulatory, and random. But it achieves the same end; they get somewhere else. Some days the orange monarchs are joined in their southward movement by long black funeral cars headed to Starksboro's Green Mountain Cemetery.

Most of the watershed's men and women did not die in California on some adventurous exploit. They died here not far from where they were born, of typhus and tuberculosis ("consumption") and apoplexy. They died of pneumonia and cancer and cholera and kidney disease. Some had scarlet fever, others were stillborn. Their headstones don't record what their death certificates do, that they succumbed to whooping cough and measles and gastritis. One or two died as "the result of a fall," some died "suddenly," and there was the occasional victim of "old age." Someone took an overdose of medicine, another drowned in Lake Champlain, and not a few died by their own hands. The suicides were men and women, young and old, farmers and carpenters and the dispossessed who took their lives at the Brattleboro insane asylum or on the town poor farms.

Today any grave opened in the watershed's cemeteries might as likely be for someone from "away"—not just the next watershed or the adjacent county, but from Massachusetts or Georgia or California—as it would be for someone born and raised within the borders of the Lewis Creek watershed. Until twenty years ago, one might count the stones in the watershed's dozen and a half cemeteries and arrive at an accurate census of all the humans since Abenaki times who were born and died here. The cemeteries tell how the land at one time cradled not just the memory, but the bones of those who had walked upon it most of their lives. The sense of place held by the living people of the watershed was embodied in the bones of those who had "gone before." Like the marl deposit at the bottom of Monkton Pond, the bones held history, keeping silent sanctuary over this fair ground.

As your eye scans the valley below from any of the quartzite bare spots on the west flank of Hogback Mountain, it eventually falls on the body of water nestled snugly against the mountain. To those who have grown up nearby, it's "The Pond"; to others a little further afield, "Bristol Pond"; and to strangers, "Winona Lake." Whatever its name, from any Hogback vantage, Sungahneetook's largest tributary pond appears as an enormous teardrop, set gently into the dolomitic valley that encircles the Hogback Ridge. Or perhaps the impression is more one of a raindrop as it distends under its own weight, plump and heavy at its earthward end, tapering to a point at its skyward end. Bristol Pond's taper runs north, to its sluggish, meandering outlet. The teardrop is surrounded by a larger, half-circle dimpling of the landscape, defined not by water but by vegetation. Corn fields and alfalfa meadows crowd in upon the pond from all but its eastern side, where there is only Hogback's quartzite shoulder, but the cultivated land yields to forest in a precise half-moon pattern around the pond. Something has kept this land in its natural crop—trees.

Walking that forest-field edge at its northern side, the lay of the land quickly reveals the reason for the pattern. A five- to ten-degree slope runs down to a narrow level margin, about the width of a cart path or grown-over woods road. The slope is all grey birch and wild cherry, with scattered swamp maples, but immediately beyond the level stretch, white cedars appear. One step taken to peel off a strip of their fibrous, twisting bark, or to crush a spray of leaves under your nose, and you are ankle-deep in water. The half-circle of trees owes its existence to this microtopography; the foot or two deep basin, though barely discernible until one is almost upon it, holds an entirely different world from the cultivated quarters that encircle it.

The half-circle basin marks the place where a huge block of glacial ice once lay, its top perhaps level with Hogback's summit. About 12,000 years ago, after the ice receded, the blue and white ice melted into itself, but not before it had dimpled the deep valley soil. Bristol Pond and its adjacent basin is a "kettle hole," the term given to such glacially created depressions. It is easy to see how early settlers would have chosen their big iron kettles as the closest analogy to these basins. Due to Hogback's presence, this kettle hole is only half complete. The center of it holds the pond, less than nine feet deep at its deepest, the southern quarter a swamp and adjacent wet meadow, bisected by a little tributary stream. The eastern margin is a diverse wetland complex, alternating between cattail marsh and sedge meadows and red maple swamp. The northern perimeter is more swamp, but between it and the pond lies a different habitat, one often associated with kettle holes—a bog.

Or is it a bog? Bogs are acidic, oxygen-poor places, deriving their moisture primarily from precipitation. They support few species of plants, and

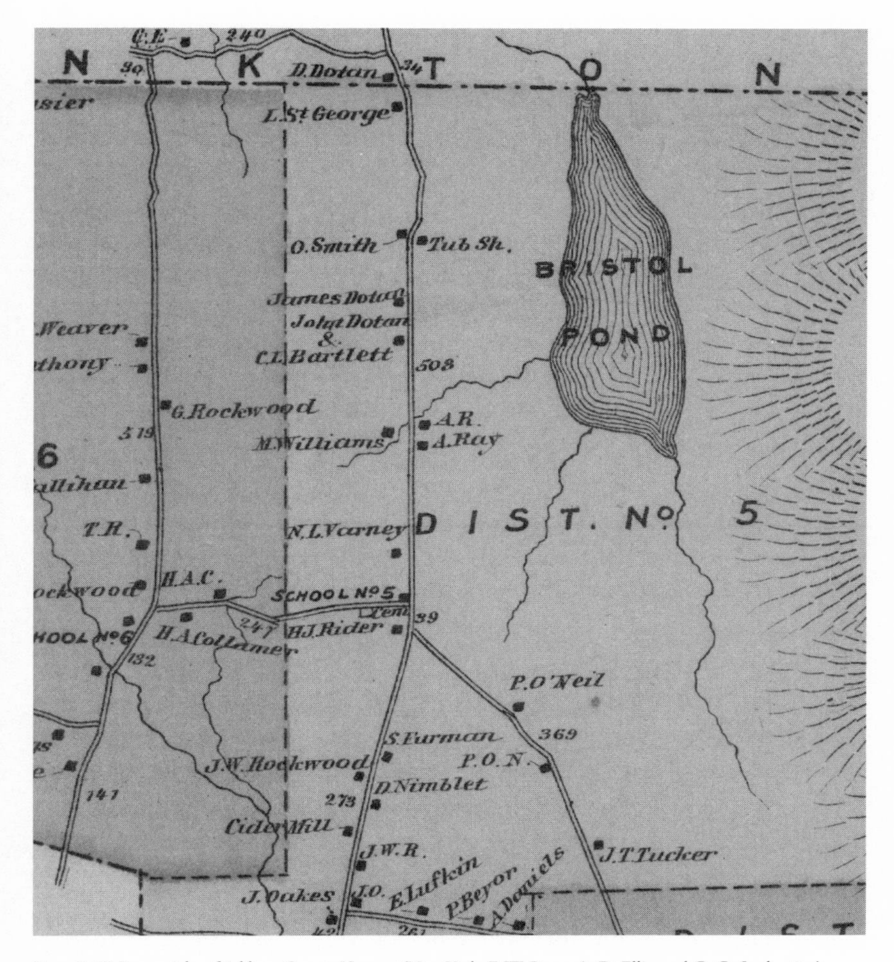

From F. W. Beers, *Atlas of Addison County, Vermont* (New York: F. W. Beers, A. D. Ellis, and G. G. Soule, 1871).

various species of *Sphagnum* moss make up the bulk of the bog's biomass. Bristol Pond "Bog" sits in a dolomitic valley, filled with calcium- and magnesium-rich gravels. The pond proper, almost 250 acres in area, is big enough so that a stiff wind whips up sizeable whitecaps, oxygenating the waters. Two or three short, steep streams tumble down through Hogback's quartzite talus into the pond, bringing more oxygen. Though there's plenty of *Sphagnum* ringing the half-water, half-land surrounding the pond, sedges predominate, and the reddish-brown peat below them is made up mostly of their incompletely decomposed remains. One foot and a thousand years below the surface the peat is fibrous, containing many recognizable plant parts. Dig down and it's like you've uncovered a grave—the parts tell unmistakably what living things once thrived there. The grey twigs, matted roots, seeds, spores, nutlets, catkins, leaves, and even faded flowers enmeshed in the peat are the bones of this plant community.

Everything, then, that this liminal landscape has to say says that it is a *fen,* not a bog. Fens are only slightly acidic to basic, mineral-rich peatlands that are open to groundwater flow, have sedges as their primary peat producer, and may support a high diversity of plants. But no one has ever called Bristol Pond Bog a fen. The farmers who run their plows right to the edge of the big depression never called it a fen, nor have the fishermen who have come to take pickerel here. Even the plant collectors who gleaned the many species of plants that grow in fens, not bogs, labeled all their collections as being from "Bristol Pond Bog."

Like so many other places in the Lewis Creek watershed, Cyrus Pringle was the first plant hunter to bring out botanical treasures from Bristol Pond Bog. His first trip there was on June 11, 1876. The previous month his rambles had taken him into the cedar swamps of the Pond Brook valley and at the foot of Mount Florona, where he had had great success, especially in securing specimens of the *Calypso* orchid. But nothing could have fully prepared him for what he found as he slogged his way over the open sedge mat on that June day. His plant press bulged with species entirely new to him, or ones that he had seen only once or twice before. There were orchids: early coral root, *Corallorhiza trifida;* hooded lady's tresses, *Spiranthes romanzoffiana;* heartleaved twayblade, *Listera cordata;* green adder's mouth, *Malaxis unifolia;* and the small round-leaved orchis, *Orchis rotundifolia*. There was the whortleberry or bog willow, *Salix pedicellaris* var. *pedicellaris,* the pitcher plant, *Sarracenia purpurea,* and the bog sedge, *Carex exilis*.

Of these and other plants, one particular species stood out amongst the dozens Pringle had collected on that long June walk—*Orchis rotundifolia*. This lovely little plant, growing in a sizable colony amongst cedar-shaded moss, had never been found in Vermont. The next day he sent a postcard to Asa Gray telling him of his discovery. Before Gray could get back to him, Charles Sprague Sargent, Director of Harvard's Botanical Garden, wrote back, saying, "I should like to get a supply of living plants for the Garden." A week later Sargent was feeling Pringle out about the "terms" Pringle might desire as a paid collector, and continued to press his new associate for more material of the little orchid from Bristol Pond Bog:

> I shall be greatly obliged if you will send me fresh specimens of your *Habenaria rotundifolia* to serve as a model for a drawing . . . now that the number of plants in cultivation is so enormous there is little hope of any horticultural success except by specialists, . . . You are sincerely to be congratulated in your courage in devoting yourself to two genera and in your taste in selecting those which you have.

Asa Gray was equally eager to get specimens of the little round-leaved orchid, for like so many members of its family, the plant was tangled in a web of taxonomic confusion. Frederick Pursh had first described the species (as *Orchis rotundifolia*) in 1814 from specimens that were collected at Hudson Bay. But in 1823, Robert Richardson, a member of Sir John Franklin's ill-fated expedition in search of a Northwest Passage, described the same plant as *Habenaria rotundifolia*. In 1835, John Lindley referred the plant to the genus *Platanthera*. *Orchis* was an apt designation, since *Orchis rotundifolia* bore the little tuberoids at the base of the stem that characterized many orchids. The Greek word for testicle, "orchis" came into use among Greek authors over 2000 years ago. They were excited and curious about the resemblance of the little tuberoids to testicles, since they lived in an age when a plant's medicinal properties were thought to be indicated by the shape of its various organs.

But many genera of orchids had testicle-shaped tuberoids—the character was far from diagnostic. To solve the tangled classification, Asa Gray needed fresh material that showed all the representative characteristics of the plant. Pringle's Bristol Pond discovery provided that. In 1877, in the *American Journal of Science,* Gray wrote:

> *Orchis rotundifolia* of Pursh, which Richardson referred to *Habenaria* (it was confidently supposed with good reason), and Lindley after him to *Platanthera,* is a genuine *Orchis,* having a pouch to the pollina-disks as manifest as that of *O. spectabilis*. This is seen in fresh flowers of the living plant sent by Mr. Pringle from Vermont to the Cambridge Botanical Garden.

Though not uncommon around Hudson Bay and in eastern Quebec, and more so west through the muskeg of Alberta and British Columbia to the Yukon and Alaska, small round-leaved orchis is one of New England's rarest plants. At the time of Pringle's discovery, it was only known from a couple of other stations in northern Aroostook County, in Maine. For Gray, who needed living or nearly living plants to study, *Orchis rotundifolia* may as well have been confined to Greenland (where it does occur in a few locations!). Asa Gray did not go to the plants; the plants came to him, and he dissected the beautiful purple-spotted flowers in the comfort of his Cambridge study. Much of *Orchis rotundifolia*'s range was still *terra incognita,* perhaps not to fur trappers, but to collecting botanists. The little colony that grew amongst the *Sphagnum* hummocks at Bristol Pond put Gray within reach of the plant. His original declaration did not go unchallenged, however. In 1880, Gray was again asking Pringle for living plants, since the British botanist Jeremy Bentham disputed Gray's claim that the species belonged in the genus *Orchis*. At the height of the orchis's flowering period, off Pringle went to Bristol Pond for plants for his mentor. Gray wrote back: "You are the most *reliable of men*. Many thanks for the two Orchis' in fine condition—with which I hope to convince Mr. Bentham that I am right."

In 1876, four days after his Bristol Pond discoveries, Pringle made his first visit to Mount Mansfield, Vermont's highest mountain. There on Mansfield's crest he added three new species to the Vermont flora, and one to that of the United States. Already wilder places were beginning to call to him. In 1878 he collected in the White Mountains, then Quebec and northern Maine. In 1879 he spent all of August in Quebec, and in 1880, a month after his jaunt to the bog to fulfill Gray's request, he made his first trip west. As forestry collector for the American Museum of Natural History under Charles Sprague Sargent, Pringle was to secure five-foot-long logs of every Western tree species. At the same time, he made collections for Asa Gray. In a sense, Sargent and Gray's confidence in Pringle had been built largely upon his success at Bristol Pond Bog. As much as his alpine work, such as his 1876 Smuggler's Notch revelations, the "bog" proved that he could ferret out rare species, preserve them carefully, and convey them quickly to Cambridge. Though he never returned there after 1880, the years between 1876 and 1880 saw him make the day-long journey to Bristol Pond at least once a year. He usually went in mid June, when most orchids and sedges were in flower, then often returned in mid to late September to gather fruit. As with so many of the places that Pringle botanized, his collections paint a fairly complete picture of the plant life of Bristol Pond Bog circa 1880. In the last century, botanists have added only about a dozen species to those which Pringle found, and it is possible that he had also collected these, but that they are not represented in herbaria. (Pringle's meticulously kept field journals, which would eliminate any doubt, have never been found.)

The only other person to come to know the flora of the Bristol Pond peatland was Ezra Brainerd of Middlebury. He and Pringle had started exchanging specimens in the early 1870s, and in October of 1877 Brainerd made his first trip to the Pringle homestead. Brainerd had become particularly interested in the family Cyperaceae, the sedges, and he pressed Pringle for as much material as possible. Receiving a bundle of sedges from Pringle in December, he wrote back, "It is rather mortifying to see that many of them were found at Bristol Bog, a place nearer us than you." The more Brainerd heard about Pringle's luck at Bristol Pond and in the cedar swamps just north of it, the more he wanted to see it himself. "I must see that swamp of Calypsos before I die! Do tell me just where it is . . . I am in despair about finding *Carex Backii*. Do you find it in more than a single locality? Can I hope to find *Orchis rotundifolia*? Let me know the day and *hour* you expect to reach Bristol Pond." Plans were laid for the two botanists to meet at the Bartlett farm, across the road from the pond, but at the last moment Brainerd's brother-in-law became ill, and he had to stay in Middlebury. Pringle sent along his *Carex* collections, though, and Brainerd worked them over thoroughly. It was not until the following June (1879) that Brainerd finally got to the bog with Pringle. Together they waded through the cedar swamp out to the bog proper, hopping between sedge hammocks and frequently dropping to their soggy knees to examine more closely some problematical *Carex* or other. The collaboration of this pair was as fine an example of botanical teamwork as ever occurred in Vermont. Brainerd was inspired and guided by the fineness of Pringle's specimens, and Pringle owed much to Brainerd's keen analytical mind, which served in the determination of so many of Pringle's critical species.

The opportunity to get to know rare members of his favorite difficult genus brought Brainerd back to collect sedges eighteen years later. By that time, a few other Vermont botanists were beginning to explore the unique plant community that flanked Bristol Pond. Willard Eggleston collected there in 1893, 1898, and 1903, and "Archie" Dike, who lived just south of the pond, visited often in the years 1895 to 1899. There was another twenty years of quiescence, and then between 1918 and 1925, D. Lewis Dutton, of Brandon, made collections there and at the "Dyke Swamp" on the other side of Hogback. In 1927, the Vermont Botanical and Bird Club took its annual summer field trip to the area, but few specimens were collected for herbaria. By that time, a conservation consciousness was beginning to emerge among amateur botanists.

Archie Dike once took a photograph of some yellow lady's slippers that he had collected at the bog—enough to fill a huge metal pail! A glance at

Willard Eggleston's field notebook reveals a similar unsettling acquisitiveness; many of the plants on his collection list are followed by numbers in parentheses—50, 60, and so on—indicating how many specimens Eggleston collected, most of them for sale to or exchange with other collectors. As he helped to fill in the botanical picture of Vermont, Eggleston was also aiding its destruction. In 1921, encouraged by the Vermont Botanical Club, the Vermont legislature passed a bill to protect rare plants, and to protect these plants from being sold for commercial purposes. When it was introduced, the bill was met with laughter from many of the legislators, but it was serious business to Vermont's amateur botanists. They debated which species should be included on the list, having very different notions of just which plants were truly rare. Of the twenty-three herbaceous species which probably belonged on the protected list, about half were plants that at one time had significant populations at Bristol Pond Bog. Just how severely overzealous collectors like Eggleston had decimated these populations is hard to tell. The most obvious pressure came from commercial collectors who'd almost extirpated plants such as ginseng *(Panax quinquefolius)* and golden seal *(Hydrastis canadensis),* but a significant impact was caused by a seemingly benign part of the Vermont populace. Elsie Kittredge, one of the Botanical Club's most active members regarding plant protection, wrote in 1923:

> The flowers of early spring I think suffer the most . . . partly because they are so eagerly sought by the children frantic to bring a bunch of posies to the school, and partly because the earth is moist and easily yielding. The later glories of wood and field and swamp are being exterminated by thoughtless "Nature Lovers," some of them having enough botanical knowledge to know better, but who give as an excuse, "I thought I might as well have them as anyone."

The farm wife who gathered armfuls of lady's slippers for a wedding decoration; the children who tore up patches of trailing arbutus to fill the kitchen with their scent; even the plant hunters who took too many specimens for sale or exchange; all lived in an age of innocence regarding plant protection. But just how many of the plants of Bristol Pond Bog were "rare," and how rare were they? Some of the rarity came about through systematic sleight of hand—that is, species splitting by certain botanists. The natural variations among populations of certain plants were given specific status by some botanists, particularly in variable genera like *Salix, Crataegus, Rubus, Carex,* and so on. To Pringle, who was working virtually alone the flora of northern Mexico, it was a bit overwhelming to see what was happening in his old stomping ground. In 1899 he wrote to Eggleston:

> I am hoping you are having as good success in the field of botany this season as you wrote of having last year—would I not have to begin anew in Vermont botany? So many new species of *Sanicula, Antennaria,* etc.! By the way let me offer a suggestion respecting our *Prenanthes*—can you not make 3 or 4 species out of the species commonest in our valley?

By this comment Pringle meant not that *Prenanthes* should be split up into three or four species, but that it was arbitrary and unwise to do the same with many of the other variable genera of the Vermont flora. Like his mentor Asa Gray, Pringle was conservative when it came to classification. His view of the plant world allowed much latitude for variation within a species—such was the order of things. The tremendous amount of time he spent collecting plants could scarcely have produced a different viewpoint, for in the field, gathering fifty or sixty examples of a single "species," he might encounter a dozen plants with anomalous characters. Such divergences might be seen by a less active botanist as indications of new species, but to Pringle they were the incipient stuff of speciation. He was content to let the plants vary without conjuring up new names for each. Another of his comments to Eggleston summed up his view: "In my struggles with synonyms I have learned it is in those floras where botanists live thickest that the greatest number of synonyms are found. It is a case of 'many men of many minds.' I like best to work among plants of wild regions, whose species have not been 'cat-hauled' about by many hands."

At the time when the plant protection law was passed, Vermont had never seen so much botanizing done within its borders, by both amateurs and professional scientists. With all this activity, many new stations had been added for certain plants previously considered "rare." Bogs, being havens for rare plants, were especially attractive to botanists, but even with all the increased bog trotting, certain plants of Bristol Bog remained rare. No one ever found another station for *Orchis rotundifolia;* there were only Bristol Bog, the Florona Mountain cedar swamp, and Dyke's Swamp across the mountain. Of the many uncommon sedges pored over by Pringle and Brainerd at Bristol Bog, three—thin-flowered sedge *(Carex tenuiflora),* Gray's pale sedge *(Carex livida var. Grayana),* and creeping sedge *(Carex chordorhiza)*—have not been found anywhere besides Bristol Bog and one or two other places in Vermont. A boreal aster, bog aster *(Aster junciformis),* is still known only from Bristol Bog, and the pretty bog willow, or "myrtle willow" as Pringle called it, was found again only once—in Colchester Bog, by A. J. Grout in 1895. These plants are undeniably rare, but as with Vermont's populations of jack pine and chinquapin oak, rarity is a relative term. It is unthinkable to a logger in Saskatchewan that jack pine could ever be a rare tree, unbelievable to a farmer in Missouri that chinquapin oak could be uncommon. Fixed as we are in place and time we forget the fluid nature of plantscapes. The whole collection of Bristol Pond's peatland flora is a truly unique assemblage—for *here, now*. Once Vermont may have been home to countless colonies of *Orchis rotundifolia,* and it may be again.

For the plant hunters from Pringle to Eggleston who walked tentatively out onto the trembling surface of the bog, this was not a consideration. They were there to hunt rare plants, not to discern the past or future of an entire plant community. In fact, the word "community" was not part of their botanical vocabulary. It was not until the first decade of the twentieth century that Vermont botanists really began to see the *forest* as well as the trees. In 1906, of a list of about 100 papers that had been presented by Botanical Club members, about three-quarters were either local floras, or new location information about Vermont plants; plant physiology was the subject of about a half dozen, and botanical methods or technique were discussed in a few. The balance were nostalgic looks at the early history of Vermont botany, focusing on the particular species added to the state flora by early botanists. Not a single paper had ecology (the study of organisms in relation to their environment) or phytosociology (the study of plant communities) as its theme. Two years later, however, H. A. Edson gave a talk on "Soil Reaction in Relation to Flora." Ecology, if only in a small way, had reached Vermont.

In 1912, George P. Burns, professor of botany at the University of Vermont, gave a paper entitled "Distribution of Bog Xerophytes" at the Botanical Club's annual winter meeting at Dartmouth College. Burns noted that many bog plants, though they grew with their roots submerged in water, had adaptations similar to those of desert plants—reduced leaf size, waxy outer surface, and so on. He also pointed out that these "xerophytic" bog plants were often found in close proximity to riparian plants having no such adaptations. "Is the bog water different from the river water?" Burns asked, in what may have been the first explicitly *ecological* talk heard by the amateur botanists gathered in Hanover. He concluded his remarks with a call for investigation: "As they grow out from the shore of our lakes and gradually fill even the deepest with an accumulation of plant remains which forms peat, turning open water into farming land, they offer an excellent opportunity for the study of plant succession and other ecological work."

Along with presenting some new questions for his fellow Vermont botanists, Burns was reinforcing some well-entrenched dogma. His statements about bog formation echo the prevailing view of botanical observers of his era—that peatlands grew out over the surface of their captive ponds, maturing through time in thickness and stability until they became suitable for trees (or in Burns's Champlain Valley–formed peatland consciousness, farmland). Burns's colleague, J. L. Hills, head of the Agricultural Experiment Station at the University of Vermont, was of a similar mind. In his report on "The Peat and Muck Deposits of Vermont" (1912), Hills said that Bristol Pond

> and its surrounding formations constitute one of Vermont's largest transitional peat bogs. It is simply a matter of time when the water now in the basin will be displaced or absorbed by the enormous amounts of moss and accumulating silt. That the pond originally presented a much larger water surface is very evident from the large swamp which surrounds it, and from the nature and depth of the deposited material. The pond is slowly turfing over.

This was the theory of Hydrarch Succession, which was becoming enormously popular at the time. Surely this sort of scenario had occurred to Pringle, Brainerd, and others, but it was an adjunct to a more pressing concern—just what plants grew in the bog. Burns and Hills, however, came of age in an era when plant communities were real entities, with developmental histories like those of an organism. They were born, grew, matured, reproduced, and then died. In espousing this successional view of vegetation, the botanist Frederick Clements, its champion, often cited the lake-to-bog-to-forest example. Even lay people with no schooling in species or succession seemed to take a similar view. One of Bristol's town historians authoritatively stated that the pond was filling in with leaves washed down from the Hogback woods.

For some reason, perhaps because they are discrete, relatively definable units of the landscape, bogs have always been a favorite template upon which to hang theories of vegetation development. The earliest account in English of vegetative dynamics is probably William King's paper in *Philosophical Transactions* in 1685; he gives an account of the origin of bog vegetation from floating mats. In 1749, a student of Linnaeus published his thesis, in which he noted the importance of *Sphagnum* in pioneering bog development. But such views implied that the world had not been created as it was once and for all time, and were silenced, or at least ignored. Clements's theory, on the other hand, came in that post-Darwinian period when "evolutionary" or "developmental" models were embraced in every field of science and social science. The concept of the plant community as an organism held great emotional and intellectual appeal, and was almost universally accepted.

In 1927, the Botanical Club gathered at Bristol Pond Bog to study plants "in their native habitats, and in their relationship to their environment." As the members collected, photographed, and sketched rare plants (for treasure hunting was still the major preoccupation), they carried with them the Clementsian view of the bog. Most had it imparted to them by Burns, and others may have even read Clements's seminal work (1916), *Plant Succession: An Analysis of the Development of Vegetation,* but all of them most likely saw the same thing as they surveyed the surroundings. Looking back toward the sloping shoulder of the basin from out on the open mat, they saw the sedge fen yield to a zone of ericaceous shrubs, then to an increasingly tall swamp forest of hemlock, cedar, white pine, and tamarack. They believed that the coniferous spires would march toward them and past them if they were to sit there and watch for enough centuries. The eventual climax in forest was as undeniable as their own deaths—such was the metaphorical power of the organismic theory.

A quarter of a century later, the theory of Hydrarch Succession had loosened its hold on observers of American plant communities. Scores of studies showed that the real world almost invariably contradicted Clements's world. Succession was no orderly progression toward climax; instead, there was a cyclic repetition of decadence, destruction, and regeneration. Any unconverted Vermont botanists need only to have joined the 1951 field trip to Bristol Pond Bog to have become believers. Thousands of bare grey snags stood where once there had been pine and tamarack. The cedar swamp, which had always been difficult to penetrate, was even more so, as there were innumerable walls of uprooted trees and shrubs. Many of the orchid colonies were completely gone, or reduced to a few individuals. The hollows between the sedge hammocks were filled with water. The previous two years had seen two catastrophes visited upon the pond—in 1949, after a century and a half of absence, beaver had recolonized Bristol Pond, and the following November, the area was hard hit by "the Great Appalachian Storm," which brought winds over seventy miles per hour down upon the drowned trees.

When Zadock Thompson first published his *Natural History of Vermont* in 1842, beaver were about as common as *Orchis rotundifolia;* in his words, "The beaver, though formerly a very common animal in Vermont, is probably now nearly or quite exterminated, none of them having been killed within the state . . . for several years." Beaver pelts were prized commodities, and since the early seventeenth century, European demand for beaver fur had decimated their populations in the northern forests of the New World. The Abenaki's neighbors to the west, the Mohawks, had become the principal instrument of the fur-trading companies, and two centuries of commerce with Europeans had radically changed their culture, along with extinguishing an animal that they had depended on for millennia. Not as dependent on the fur trade, the Abenaki continued to trap and hunt *t' mah queh*—beaver—for their own needs, until white settlers drove them and the beaver from the Champlain Valley and the rest of Vermont. In a century, whites did what 10,000 years of native hunting failed to do—exterminate the great rodent of the north woods.

In fact, destruction of the beaver may have provided the niche for *Orchis rotundifolia*. There had never before been a time in Vermont when there were people, but no beaver. Beaver had always been periodically flooding the fens, bogs, and cedar swamps of the Champlain Valley, and even if there had been an Abenaki herbalist looking for the little orchid for some magical or curative need, he may not have been able to find it. In the last five thousand years, perhaps it was only in that brief century and a half, between 1800 and 1950, when *t' mah queh* was absent, that the rare orchid flourished at Bristol Pond. Few mourned its passing; though many lamented the arrival of the disruptive beaver, it was because he robbed valuable farmland, not because he flooded out orchid habitat. At Bristol Pond, his reappearance was even noted with some pleasure, since many believed that the legendary snake population of the pond had decreased some since the raising of the water level.

Botanists always called the peatland around Bristol Pond a "bog," though it was not a bog: The people who have written town histories of Bristol stated unequivocally that the pond was being filled in by leaves and other organic matter being washed into it, though this is hardly the case. Learned professors pronounced that it was just a matter of time before the pond filled in, yet this is not so; townspeople for over a century have avoided swimming in the pond because of the snakes, but the stories diverge greatly from the facts. A rare plant seemingly owed its existence at the pond to the destructive ways of transplanted Europeans. Even the name of the placid body of water is illusive and contradictory. In the 1960s, tourists cruising south to Ripton to see Robert Frost's old stomping ground could glance at their highway map and see the teardrop-shaped pond called Lake Winona. With such a picturesque name, some may have been inspired to make a detour to have a picnic lunch there. Who knows how many innocent picnickers drove down the access road to be disappointed by Lake Winona's reality—a sea of drowned trees, tea-colored water rank with pond weeds, and the occasional whiff of "marsh gas"—methane. Not just the gasoline companies promoted the name "Lake Winona"—U.S. Geological Survey maps, Vermont Highway Department maps, and recently published atlases read the same way. The cartographers who laid the letters down across the blue spot on the map never questioned its correctness. Only a linguist or geographer would have wondered why an Iroquois name came to be applied to a pond in the heart of Abenaki country.

The name has nothing to do with Indian legends or ancient cultural contact, but everything to do with myth making and storytelling. Bordering Bristol Pond there is a farm along the Monkton Road that once belonged to Truman Crane Varney, one of Bristol's outstanding citizens. Born and schooled there, he was Master of the Bristol Grange for seven years, a selectman, and Bristol's representative to the state legislature. By the time he was elected to this office for the second time in 1930, he had retired from the farm to live in the village near his own Methodist Church. Living with him and his wife were his grandchildren, Charles and Winona; their mother, Vera Varney Meilleur, had died after a long illness in 1926. When their grandfather enrolled them in the village school, he anglicized the children's French surname to Mayer; it was still an era when the good Quaker Varney stock was not eager to mix with French-Canadians. The little girl Winona was the apple of her grandfather's eye, and in 1931, when the legislative session began in Montpelier, the Varneys took Winona with them. There they shared a house with state senator Clarence Lathrop and his family. Perhaps Senator Lathrop, the more experienced legislator, suggested how Varney might realize one of his aims for the 1931 session—to rename Bristol Pond after his granddaughter. In spite of a previously adopted policy of the legislature to reject any bill renaming roads, lakes, and mountains, Varney's proposed bill met with little opposition. The committee reporting on the bill said that the new name would be "a means in aiding in advertising the lake." Despite such a fantastic claim, the bill was accepted by the House and passed on to the Senate, where Senator Lathrop used his influence to quell the brief and limited opposition it met there. On February 17, 1931, the bill became law: "The pond situated in the town of Bristol commonly called Bristol Pond, is hereby named and designated as Winona Lake." Only a week later, a less trivial bill—"An act for human betterment via voluntary sterilization"—made its first appearance before the Senate. Truman Crane Varney was one of 145 representatives who voted in favor of passage. Little did he know that his name was on one of the Eugenics Survey pedigree charts, and that Winona Mayer was one of the pupils that eugenics fieldworker Martha Wadman was so interested in when she wrote to the principal of the Bristol school for information concerning the intelligence of Varney offspring.

Winona Mayer graduated five years later from Bristol High School, salutatorian of her class. In 1941, when her grandfather died, she was serving as a nurse at a war-time army hospital. Her brother Charles was a second lieutenant fighting the Nazis in Europe. In their grandfather's obituary notice they were "Meilleur," not Mayer. Their true identity seemed ensured, but not so the identity of the pond they had known and loved while growing up on their grandfather's farm. Though the legislature said "Winona Lake," the maps said "Lake Winona." The shape of the pond probably dictated the inversion; "Winona" was easier to fit at the pond's broad midsection than at its constricted northern end. In 1965, a federal agency got into the act in an attempt to end the confusion; the United States Board of Geographic Names condemned Bristol Pond and Lake Winona, settling on the name pronounced in the 1931 bill. Neither cartographers nor townspeople have taken any note.

By association, the story of the pond's name runs much deeper. Vera Varney Meilleur loved Longfellow, as did her father, and her father's father before him. His *Song of Hiawatha* reinforced their notions of individual action, heroism, and manifest destiny. It was from that poem that Vera chose the name of her first child—Winona. Longfellow's "Wenonah," a "tall and slender" Indian maiden "with the beauty of the moonlight," was borrowed from the Ojibwa Indians on the southern shore of Lake Superior, while the rest of the material he plucked haphazardly from a number of mid-nineteenth-century sources. The main character of the poem, Hiawatha, was really an Onondaga chief who was instrumental in the founding of the League of the Iroquois, not an Algonquian as portrayed by Longfellow.

(One informed writer made this comment twenty-five years after the publication of Longfellow's famous poem: "If a Chinese traveler, during the middle ages, inquiring into the history and religion of the western nations, had confounded King Alfred with King Arthur, and both with Odin, he would not have made a more preposterous confusion of names and characters.") Continuing the confusion, Longfellow wed Wenonah and her misidentified kinfolk to the legendary traditions of northern Europe by borrowing the meter and spirit of his poem from the *Kalevala,* an ancient Finnish epic poem. The *Kalevala* itself had its roots in still older story cycles that stretched back to the beginning of human habitation of Europe. All this borrowing and mistelling was nothing new, and for an audience starving for myth, as Longfellow's antebellum America was, it certainly did not matter. What was wanted, as always, was a good story.

In the *Kalevala,* after recounting the creation of Otso, the bear, the hero of the poem tells how the bear was nurtured by Mielikki, the forest's mistress:

> There she rocked the charming object
> And she rocked the lovely creature
> Underneath a spreading fir-tree,
> Underneath a blooming pine-tree.
> Thus it was the bear was nurtured,
> And the furry beast was fostered,
> There beside a bush of honey,
> In a forest dripping honey.

Otso swore to Meilikki that if she granted him sharp claws and strong teeth he would never use them for evil or mischief, and in the sagas that follow the *Kalevala,* the bear keeps his word and protects the people of the North. In return, he is treated by them with awe and respect.

In Longfellow's hands the bear is no longer a heroic creature, but a beast to be subdued by man. Hiawatha's father, Mudjekeewis, overpowers the "great bear of the mountains" by cunning, and steals a belt of wampum from him. Mudjekeewis utters no prayer for the bear, does not welcome the great forest creature. The bear is simply killed.

> With the heavy blow bewildered
> Rose the great bear of the mountains;
> But his knees beneath him trembled,
> And he whispered like a woman.

The path along which the symbol of the noble bear had come was as convoluted as the pond's name. Out of the northern European woods the bear had lumbered into the imagination of Neanderthal human beings 50,000 years ago. To him the bear was an animal of great power—somehow he disappeared into the underworld each fall, only to return as the earth grew green again. In some of the caves where bears escaped winter, Neanderthal man groped dimly toward his own evolving humanity. He left handprints on the walls of the caves, perhaps in imitation of the claw marks left on trees by his cousin cave bears. More than the horse, bison, deer, or mammoth, the bear *was* his cousin—omnivorous, upright, curious, playful, and intelligent. As northern Europe shed its primal past, myth became literature, ritual became art. The sacred bear of the cave became the secular bear of the *Kalevala*. Humanity ascending, bear descending.

Vera Meilleur had no Neanderthal thoughts as she learned to recite her father and grandfather's favorite poem. The words rolled from her lips—"By the shores of Gitchigumee, By the shining Big Sea water . . . "—in a tradition vastly different from the one that lead to the *Kalevala*. Yet there was power in the incantation still, and who could deny the poetry of the Indian maiden's name—"Wenonah"? That Vera's first-born was named "Winona" was no great surprise, and that Winona's grandfather should have immortalized a cherished memory onto the landscape seems natural also. The great irony is that the pond's romantic name came so long and tortuous a route, from Europe's Paleolithic to Finland's Middle Ages to Longfellow's study at Harvard in 1854 to a farmhouse near the shore of a kettle-hole pond. Some more suitable name might have come directly from the pond's own first people—Paleo-Indians. When the first Paleo-Indian bands hunted along the margin of the pond, not just the "bog" but much of the entire landscape was treeless. There were patches of open spruce parklands, scattered colonies of birch and aspen, and stands of fir, spruce, and tamarack along the waterways, but most of the land was lichen-covered tundra. Less than five miles to the west, waves broke on the sandy shoreline of a sea that stretched to near present-day Lake Ontario, and south to near where Whitehall, New York, lies today. In that 20,000-square-mile inland "Champlain Sea" swam cold-water fish like capelin and sculpin and a variety of marine mammals. There were beluga, finback, and bowhead whales; ringed, harp, and bearded seals. Gyrfalcon and snowy owls hunted lemming and snowshoe hares in the valley surrounding Hogback. Mammoth, mastodon, woodland musk-ox, moose-elk, and caribou grazed tundra herbs where today Holsteins munch clover and alfalfa.

The tundra plants left their records in the peatland at Bristol Pond and in other places, the ebb and flow of vegetation chronicled by the pollen laid down in the bogs. The animals we know from their bones; closest to Bristol Pond, the Ice Age mammal archives include the woolly mammoth from a Mount Holly muck bed (discovered in 1848, beaver-gnawed wood was found eleven feet below the surface with the hairy proboscidean's remains, confirming *t' mah queh*'s presence in Pleistocene peatlands) and the beluga that beached in a Champlain Sea estuary in Charlotte. But what of the people who walked that 11,000-year-old landscape? What did they leave behind?

At Bristol Pond, they left three signs of their presence, or, more correctly, that is all we have found of them. In the 1920s, James Manley of Milton, Vermont, collected three unique spear points from the west side of the pond. When he found them, the term "Paleo-Indian" had not yet been coined, and the curious stone weapons found their place in his artifact collection quite anonymously. There was no elaborate taxonomy of projectile points then, and hardly any notion that people might have been in New England more than 10,000 years before. It was not until the 1930s that the Paleo-Indians of North America were given a name, but at that point they were imagined by Western archaeologists. The diligent Vermont collectors, who spent their leisure hours arrowhead hunting, but rarely if ever read *American Antiquity* and other archaeological journals, were for the most part unaware of the significance of their finds.

The three stone weapons collected by Manley are "fluted points," so called because they have on each side a long central "flute," or channel. These channels facilitated hafting the point onto a spear or lance. Two of the fluted points are of that ubiquitous stone tool material, Cheshire quartzite, while another is made of a reddish-brown chert, commonly known as "Colchester jasper." That the quarry source of this point lies some twenty-five miles away is not surprising, since the Paleo-Indians who loosed the spears holding these three points moved over large areas, to seasonal camps where they could intercept and attack migratory herds of caribou and other big game animals. They lived in small bands and traveled light, so that they could exploit a variety of resources. The same people who cast their spears at Bristol Pond may have clubbed harp seals on the pack ice of the Champlain Sea, harpooned beluga whales in shallow estuaries near the shore, and they may have confronted huddled circles of musk-oxen in dozens of places in the valley between the Champlain Sea and the Green Mountains. Opportunists, they also took beaver in their lodges, caught fish and birds, and gathered wild berries, greens, and roots.

From the time of Paleo-Indian arrival in Vermont, Bristol Pond was likely a favored hunting place. Caribou, mastodon, mammoth, and other large game mammals would have watered there, and they may have been caught by the unsure footing of the peatlands surrounding the pond. If peat mining had ever caught on in Vermont, scores of Ice Age mammal bones could have been extracted from there, some perhaps with fluted points of Cheshire quartzite mingled among them. The quartzite bluffs above the pond would have been perfect places from which to survey the parkland-tundra landscape of the Pond Brook valley. Camped there, the Bristol Pond Paleo-Indians would often come upon trees mauled by bears, or they may have stumbled on their dens in the quartzite talus or at the base of trees along the Hogback Ridge. In winter the small bands of people told stories about bears, tales of man's kinship with the bear, and of the dance between human and animal ways. The stories affirmed that bears, masters of food getting, taught the people how to survive by their varied and relentless pursuit of all things edible. In a deeper way, the bear's passage into the earth and reemergence with the first uncurling leaves of skunk cabbage and sprouting sedges taught that even death was endured.

To help his people communicate with the animal gods, the shaman of the little group at Bristol Pond sometimes *became* the bear. Before a roaring fire he leapt and lumbered in rhythm with the beating of a skin drum, summoning his bear helper. Above the ritual blinked the bear's celestial counterparts—the signal constellations of the north, Ursa Major and Ursa Minor, comprising together the "Great Bear" of the heavens. Sometimes as it slept the bear was killed by Bristol Pond's first people. Taken from some snow crypt in the grey talus, its entrails healed the wounds of winter; its stored fat was a marvelous ointment, its flesh nourishment amid winter's scarcity.

The long winter of the Pleistocene faded even further for the Paleo-Indians of Bristol Pond around 7500 B.C. Fossil pollen shows that the climate grew warm; spruce and fir declined, while white pine and various species of oak were on the rise. Forest diversity and productivity decreased, and though the woods supported red fox, marten, wolverine, lynx, and other animals, the large Ice Age mammals that earlier generations were used to hunting were becoming scarce or extinct. The Champlain Sea, cut off from the Atlantic by the rebound of the ice-free land, was rapidly flushed of its salt water, and became a cold, sterile lake for awhile. The extensive wetlands that border present-day Lake Champlain did not form until after 6000 B.C. Until recently, the consensus had always been that Paleo-Indians retreated northeastward toward the coast, leaving Vermont empty between 7000 B.C. and 5000 B.C. But it seems that that hiatus may be more a product of our limited knowledge than actual fact.

The era of prehistoric occupation that followed the Paleo-Indian period is referred to as the Archaic, and its earlier manifestation in Vermont, known as the "Vergennes Archaic," comes from a cluster of archaeological sites only an Archaic hunter's day's ramble from Bristol Pond. The Archaic hunter sought smaller quarry than his Paleo-Indian predecessor—moose, deer, beaver, muskrat, fox, marten, otter, and so on. In the marshes of Bristol Pond he hunted ducks and geese. He plied the waters of the pond in dugout canoes, spearing fish and turtles from the tea-colored water and hunting gulls and loons in the middle of the kettle pond. He gathered acorns, beechnuts, and chestnuts from the dry forests on Hogback's western slope, competed with bears for ripening blueberries and huckleberries on the "bare spots"—the naked quartzite ledges that were furrowed with polypody fern and *Woodsia*. In the swamp at the north end of Bristol Pond, he searched for branches of ironwood to make his spearthrower, or "atlatl" as the archaeologists call it. His spears were usually tipped with large lanceolate points of Cheshire quartzite quarried along the base of Hogback.

Archaic humans left relatively more evidence of their presence around Bristol Pond than their Paleo-Indian antecedents. Scores of "Otter Creek" points, the chipped stone points diagnostic of the Archaic, cover the nearby fields. "Bannerstones" and "birdstones"—both names given to differently shaped atlatl weights—have also been found nearby. The quartzite quarries that rang with the sound of stone for over one hundred generations of the Archaic period are mute now, but the deep piles of chipping debris speak of their past use. What of those many generations? Did they leave behind more than stones to puzzle over? Where are their *bones*?

Many of their bones lie a few feet below the roots of the corn that grows each year in the rolling field just north of the Bristol Pond "basin." A hundred years ago, this was Henry Williams's field, the best of those that surrounded the big two-story frame house. He knew that he had settled on land occupied by "ancient red men," as they were so often called, because each year, along with the "hard heads"—pieces of glacially-strewn quartzite—his plow turned up dozens of arrowheads, spear points, hammerstones, and pottery shards. The big quartzite bluff across the road from his farmhouse was littered with artifacts, and he kept finds from there and from his annual plow harvest neatly arranged in boxes in the barn.

The big field between the house and the pond was on a gentle rolling knoll of sand, part of the large kame terrace that flanked Bristol Pond. This deposit of well-sorted sands and gravels had been formed by streams that ran through channels on the retreating Pleistocene ice mass, then collapsed on to the underlying landscape after the glacier had melted. Around 1880, the town of Bristol began quarrying the knoll for material to help build a nearby dam. The workmen may have noticed the series of low mounds scattered through the field, and their meaning must have become clear as they dug into the earth. Bones spilled out of the earth everywhere they dug. Hamilton Child, reporting on the burial ground in his 1882 *Gazetteer of Addison County,* said that the "bones were found so numerous that they were obliged to desist from their labor and procure gravel elsewhere."

Looking back on the discovery a century later, just whose bones came tumbling out of the gravel is hard to say. Like most nineteenth-century archaeological finds, the accounts are confused and contradictory. The report closest in time to the discovery was made by Harvey Munsill of Bristol. In an undated manuscript history of Bristol he says: "Something more than a year ago an Indian burial ground was discovered. Four or five skeletons were found in sitting position in a sand bank." By 1913, when the Vermont Bureau of Publicity's "Vermont: Land of Green Mountains" brochure came out, Williams's entire archaeological collection seemed to have come out of the graves: "Skeletons, arrowheads, war clubs, stone knives and hammers, and fragments of pottery were unearthed."

The best account was given in 1898 by F. H. Williams, who made a "scientific" investigation of the ancient cemetery:

> Digging down about two feet through soil that showed plainly marks of previous disturbance, we came to a level floor made of round cobble-stones, perhaps three feet long by two feet wide. When these stones were removed we found yet another layer, beneath which showed plain evidence of severe heating. Between the two layers of stones was an inch or more of charcoal. The lower floor rested on undisturbed gravel.

An opportunity for more understanding of the burial ground came in 1937, when William Ross and John Bailey of the Champlain Valley Archaeological Society revisited the site. They were unable to locate any graves.

Culling the credible parts of the various accounts leaves the following portrait of the burials: They were congregated in a special place reserved for interment of the dead, not adjacent to a village site; Munsill's "sitting position" most likely translates to a tightly flexed burial position; the many skeletons suggest that cremation was uncommon or absent; no artifacts are known with any certainty as coming from the graves, but a bannerstone was supposed to have been found with one of the burials. All of these facts—the flexed skeletons, low artificial mounds, the bannerstone, plus the inferred negative evidence of the absence of pottery, which would indicate a later, Early Woodland, cemetery—very tentatively point to a Late Archaic date for the burials.

That we can hint at a date and give a name to a period of prehistory is a great improvement over the speculation of a century ago. For a taste of how imaginative such speculation was, we can turn to the master of the Red Sandrock, John Bulkley Perry. Perry was the first person to write up an account (in 1868) of an ancient cemetery, known as the "Hemp yard" or "Frink cemetery," which was located in Swanton on a high sandy ridge above the Missisquoi River. (The site had once been a Champlain Sea beach.) Aware of the seventeenth-century Abenaki village at *Masipskoik* (i.e., "Mississquoi"), Perry conjectured that the site had served people before the "St. Francis Indians." He believed the cemetery to have "belonged to a race long since extinct, a race which inhabited the country before either the Iroquois or the Algonquins."

Perry had his own notions, not very different from many of his scientific contemporaries, of who the ancient people were whose bones were interred in the Swanton sand:

> From these graves I have collected pieces of earthen ware, adorned with . . . hieroglyphics of undoubted antiquity, and which to my mind [are] . . . unmistakeable evidence . . . of a people closely allied in their sentiments and habits to the nations of the East. Reference is now more particularly made to earthen tubes, somewhat in the shape of a flute or pipe . . . ornamented with hieroglyphics of a moral or religious character. The markings so far as I can make them out, are closely akin to those employed as well in the Eleusinian rites, as in the old Cyribaic mysteries of Samothrace.

Perry and his generation could not comprehend that a thoughtful, skilled group of people had sprung from the very soil upon which he preached. That the Missisquoi delta and Bristol Pond and the rest of the valley lying between the two mountain ranges nurtured a people more ancient than the civilizations of the Old World never entered his mind. The period of this lost race's existence was suggested by artifacts found with the burials:

> Among these remains are also specimens which might seem at once to hint at the Noachian deluge, and to symbolize the deliverance from it. A canoe, with what appears to be a bird, perhaps a dove, wrought in stone, is one of the emblems referred to. This, when compared with some of the Mexican antiquities interpreted as having such a signification, seems certainly with as much clearness as they, to point to the flood associated with the name of Noah.

Lastly, Perry speculated about the physical characteristics of the "lost race," since an old man had once told him about finding a gigantic skeleton while building a road forty years earlier in Swanton. The bones had been examined by a local doctor, who pronounced that they belonged to an individual seven and a half to eight feet in height. Not only were they giants physically, believed Perry, but their technological skill loomed larger than that of the region's more recent aboriginal occupants. "In cultivation and refinement," said Perry, "they were much further advanced than the later inhabitants of the forest." By this he meant the "St. Francis Indians," whom he knew, or more exactly, knew of, in the same way that most mid-nineteenth century Vermonters knew the Abenaki—incompletely and incorrectly.

The Abenaki were for Perry almost as distant as the people who had carved images in their smoking pipes and made their spearthrower weights in the shape of birds and turtles. In his 1860 history of Swanton, Perry concluded:

> Thus ends the account of the St. Francis Indians—the remnant of a great tribe, and a very powerful nation of Red Men, who played a conspicuous part in the early history of New England; and who were by no means without prominence in the wild and stirring times enacted at an early day in Vermont. . . . But most of these people have disappeared; their bloom has faded, their strength is wasted; they are now few, in the thin and yellow leaf of Autumn; and soon they will be no more. Not long hence there will be "the last" of the Abenaquis, as there has been "of the Mahicans."

Indeed, in his report on the Swanton cemetery, Perry cites "one of the few surviving members of the tribe," as if the last yellow leaf were about to fall and disappear into the forest mould.

But the Abenakis in Vermont did not disappear. While Perry wrote his epistles about eclipsed peoples, several hundred Abenakis were his invisible neighbors. Some lived in Swanton's Back Bay community, and others in liminal places like the floodplain woods of the Missisquoi delta. Though no such Abenaki community lingered around Bristol Pond, the native people were nowhere near as distant as the kame terrace burials in Henry Williams's cornfield seemed to suggest. In the early 1860s, Harvey Munsill heard that two or three families of Indians had made a camp in the woods "a little north of Bristol Village, between the road and the mountain east." These were probably families from Odanak (St. Francis) who had come by way of the St. Lawrence, the Richelieu, Lake Champlain, and then Little Otter in their birch bark canoes to their ancestral hunting, fishing, and trapping grounds around Bristol Pond and Hogback. When Munsill and two friends stopped to visit with the Abenakis to learn about their ways, among them was a man who pronounced himself to be about 98 years old, and who said he had been born on the east side of Bristol Pond. In his broken English, he answered questions about the relics that paved the soil surrounding the pond. The aged Abenaki said that he had never used a stone arrowhead, and though he had heard of some who used them, he himself had never seen one used. Even in his youth he had always hunted with iron arrowpoints made in a shape similar to the stone points.

Munsill never asked, or at least never recorded, the old man's name for his birthplace, the kettle-hole pond that had seen countless generations born and die near its shore. What did they call it? What name might the Archaic people laid to rest in the kame sand have called it, and the Paleo-Indian hunters who came before them? The best way to get at such a name was via the Abenaki, whose language had roots in the Archaic. Though pushed back to Odanak or shacks in lakeside floodplain woods, Bristol Pond and "the Mountain" were still vital places for the Abenaki in the nineteenth century. Rowland Robinson recorded an encounter on April 30, 1881, that hints at how vital:

> I went today to see my Indian friends, Joe Tocksoose, Louis Tahmont, his wife, a pretty and neat ⅛ Indian woman, and a little brown baby, a newcomer since my last visit—a quiet little soul, not yet making much noise in the world. Tahmont says they do not strap their babies on a board, in the old fashion—Another instance of the little they have gained by civilization, for there can be no more convenient disposal of a baby than lashing it to a board and hanging it up between times, such as feeding, spanking, etc. . . . They had been on Tuesday last, to Hog's Back for canoe bark and got it. Tocksoose said they found the snow (here [at Ferrisburg] the grass was green and almost the last drift gone) up to their knees, and camped on the

mountain over night. They had everything ready for the building of the canoes: bark, cedar for ribs and raves, but the spruce roots for sewing. These they could not get on the Mt. for the frozen ground, but expect to get on Dead Creek. If they cannot get them, will use second growth White Oak splints, *not* from the roots. The man of the family has a canoe begun, which Tocksoose took me to see. . . . Tocksoose was in the Union Army, a N.Y. Reg[iment]. Tahmont and he have seen Thoreau's Joe Polis—"Tahmont Swasin" of the "Maine Woods" was undoubtedly Louis' brother Swasin Tahmont, now gone by the "Strong Water Stream" to happier hunting grounds than these. . . . Their language is beautiful to hear; like the gurgle of a stream. They might talk a half hour steadily, and the brown baby sleep undisturbed as if asleep by a brook, but when my harsh English broke in the young Abenaki would awake.

Though Robinson recorded many Abenaki place names, he failed to record one for Bristol Pond. What might Tocksoose and Tahmont have called the tear-shaped pond as they looked down on it from their camping place on Hogback? From there they had an eagle's eye view of the land traversed by their ancestors for ten thousand years. They could see clearly where Henry Williams's fields yielded to the swamp woods of cedar and red maple and elm and white pine, dotted here and there with little groves of tamarack. They saw the treeless sedge fen between the cedar swamp and the tea-colored water of the pond. They could make out the meandering path of open water through the sedge-cattail meadows of the northern end, and imagine it finally becoming an outlet stream through the tiny ravine whose wooded banks cut one of the cornfields. Perhaps they saw the opening in the land where the gravel pit graves were uncovered.

Seen from up on Hogback, the pond seemed to struggle to empty itself. It hung precariously between being a true kettle-hole pond—with no outlets—and an outlying branch of Sungahneetuk, connected by the small north-running brook to the larger watershed. Perhaps this character, of nearly emptying into itself, rather than running away, might have given the pond its aboriginal name. Henry Thoreau recorded just such a name from Louis Tahmont's brother, Swasin (a modification of the French Catholic baptismal name—Joachim—bestowed on him at the Saint Francis mission), when on his journey through the Maine woods in 1853 he asked the name of Moosehead Lake:

Joe [Polis, Thoreau's Penobscot Indian guide] answered, *Sebamook;* Tahmunt pronounced it *Sebemook.* When I asked what it meant, they answered, "Moosehead Lake." At length, getting my meaning, they alternately

> repeated the word over to themselves, as a philologist might,—*Sebamook,*—*Sebamook,*—now and then comparing notes in Indian; for there was a slight difference in their dialects; and finally Tahmunt said , "Ugh, I know,"—and he rose up partly on the moosehide,—"like as here is a place, and there is a place," pointing to different parts of the hide, "and you take water from there and fill this, and it stays here; that is *Sebamook.*" I understood him to mean that it was a reservoir of water which did not run away.

"Sebamook"—the sound of it would have pleased Rowland Robinson had he heard it. It did not take in the shallowness of the pond, nor the tannin color of the water; it didn't encompass the name of the old man born on the east shore, perhaps the last Abenaki born there, and the last to die there also (he died shortly after Munsill talked to him, and was buried in the Greenwood Cemetery, dug into glacial kame gravel west of Bristol Village); its syllables didn't speak of beaver and round-leaved orchids, nor of bog versus fen; "Sebamook" did not record Winona Meilleur or Longfellow or the *Kalevala,* nor the name of Joe Tocksoose's clan—Awahsoose, the Bear. It spoke not of ancient burials or long-told stories of when there were giants—giant men, giant beaver, giant elk and huge woolly creatures whose roar shook the snow from the tops of spruce trees. "You take water from there and fill this, and it stays here; that is *Sebamook.*"

What names will future generations call Bristol Pond, Monkton Pond, Lewis Creek? At this moment, ten generations removed from the first English and French settlers who coined these names, we still find satisfactory names whose associations lie far across the Atlantic Ocean. "Lewis Creek" is known to more Americans today than ever, not because of the celebrity of the Champlain Valley stream, but because its name has been borrowed for a line of sportswear. One hundred years hence, this association may be more familiar to those who cross the Creek on Route 7 (just above the falls that gave it its Abenaki name) than its intended commemoration of a French King. Whatever future names are bestowed on these bodies of water, we can take heart that they will speak of their time as surely as the current names speak of our own.

CHAPTER 10

Blood Streams

> *The outstanding lesson of the study that has been made of selected defective families is this: . . . "blood has told," and will keep right on "telling" in future generations. "Running water purifies itself." The stream of germ-plasm does not seem to.*
> —Henry Farnham Perkins

EWIS CREEK really begins to take shape in Starksboro, on the schistose Front Range hills where the creek most resembles a tree in its form. East of the last north–south trending dolomitic valley the headwater streams branch into a typical dendritic pattern. First-order streams—ephemeral, often spring-fed, faintly discernible channels that become thin cobble-stone lines of nettle and maidenhair by summer's end—join with others of their order at anonymous junctions marked only by a slight increase in volume. Second-order segments run off from these meetings to meet their morphometric equivalents and yield third-order offspring. In March and April, as the day lengthens and the sun climbs higher, as the rock maples imbibe the precious sap stored during winter's sleep in their ever-branching roots, these anonymous trickles turn into torrents, tearing away soil until they tumble hundred-foot hemlocks and hurl great angular pieces of the mountainside down their raging lengths. This stream that finds its outlet at Lake Champlain at an elevation of 90 feet takes shape on the Starksboro hills at elevations up to 2500 feet, where snow lies deep weeks after the last drifts behind even north-facing knolls in the Champlain Valley have vanished. The culverts under the gravel roads out there are six feet wide, comical in July

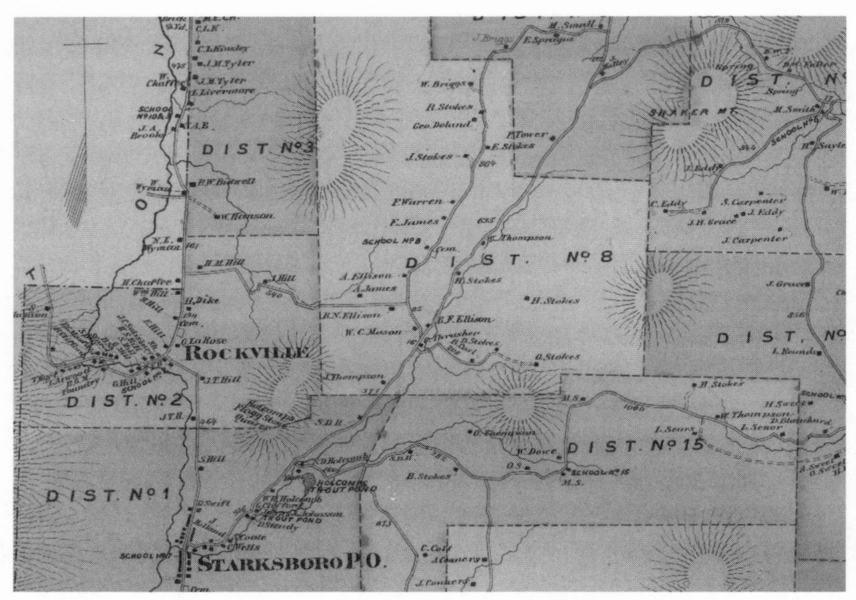

From F. W. Beers, *Atlas of Addison County, Vermont* (New York: F. W. Beers, A. D. Ellis, and G. G. Soule, 1871).

when a piddling flow spills out of them, but practical when snow-melt half fills them only three months before.

It seems hardly possible that these wild waters will end up placid and warm in the clay-banked channel that meanders through Charlotte and Ferrisburg. Less likely seems the fact that the steep, dark hemlock and yellow birch ravines here belong to the same body of water that falls only twenty feet between the big intervale below the "Hollow" and its outlet at Hawkins Bay—a distance of perhaps five miles. In the ravine below the Butler Cemetery in Starksboro, the Creek falls 200 feet over a distance of less than 800 feet, which equates to a drop of three inches for every foot the stream travels. A half-dozen similar ravines unite in a branch of Lewis Creek that splits one of Starksboro's hill communities—Hillsboro—in two.

When the first deed to land in this part of the watershed was given in 1798 to Samuel Hill of Barrington, New Hampshire, few white men had walked these uppermost branches of Lewis Creek, spring or fall. When Samuel began clearing his land, the nearest neighbor was three miles away. A year later his brother John purchased land near the "Twin Bridges," where two unnamed second-order streams join in a little gravel-bottomed pool that has been dammed up by generation after generation of boys to make a fine swimming hole. By 1800, three more Hill brothers had left the thickly settled lands of New Hampshire's Piscataqua Valley, whose tall pines had long since been turned into masts and spars for the ships of the British Navy. They came from a land of isinglass and old Indian clearings, where farming had been going on for over a century and a half, to stony hills cleared only occasionally by lightning fires or hurricanes. What few Indians had walked the upper reaches of Sungahneetook would never have thought of planting corn there.

"Stark"—the half dozen dictionary meanings may never have occurred to the Hill brothers as descriptive of themselves or their new land. "1. Strongly constructed; sturdy; stout; robust; vigorous"; Samuel, John, William, Thomas, and Lemuel were all of this stock. "2. Lacking in flexibility"; the Hills already showed their flexibility by transplanting themselves from a place where you could smell the salt in the coastal fog to hills often locked in chilly snow-laden clouds. "3. Pure; sheer; utter"; there was much of their new lives that must have seemed so, stripped down to essentials. "4. Violently stormy or windy; inclement"; yes, the south wind blew much more fiercely up on their new farms than it did along the foot of Hogback Mountain. "5. Bleak; barren; desolate; bare; empty"; though full of wildness, the deep woods they cleared appeared empty to men used to village greens and turnpikes. "6. Sharply delineated"; like the rock crags or the Adirondack horizon onto which their new farms gazed, life was stark—sharply delineated—here. There were few in-betweens. Gray was the color of beech trunks and the sky in November and their hair before too very long, but not of their thoughts. These were etched in their minds like glacial striations on the schist ledges that dotted their farms.

In 1805, Starksboro chose Samuel Hill as its second representative to the state legislature. In 1810, the last of the six Hill brothers, Francis, bought land across the steep valley. The hills that shed their water to form the finger-tip channels of Lewis Creek became "Hillsboro," for the family who had settled them. Hillsboro had a life of its own, separated as it was from the mills and farms of the town's principal valley. In 1817, an itinerant Baptist minister began holding church services in the big two-story farmhouses of the Hill families. Roads were built that connected Hillsboro to Big Hollow to the north and South Starksboro to the south, and to the families there and beyond—the Browns, Places, Emmonses. Like the Bible advised, the Hills were fruitful and multiplied.

By 1840, the Hills were leaving Hillsboro, this time to farm the more fertile bottom lands flanking Lewis Creek. In their wake came new immigrants—Irish families dispossessed of everything by the Potato Plague. While half a million Irish men, women, and children poured into New England's mill towns to work at cotton and woolen factories, tanneries, and on the railroads, a few who still believed agriculture could sustain them took jobs as hired hands on farms. Thomas Hannan, from County Limerick, was the first to come to Starksboro, in 1848. By 1852, he had achieved something he might never have hoped for in his homeland—his own farm. Thomas Casey, a stone mason from County Clare, had equal good fortune, taking a mortgage on fifty acres in Hillsboro for $300. In a few years, there followed Timothy Butler, John O'Connor, John Welch, Thomas Dillon, John Fitzgerald, Patrick Leonard, Daniel Hayes, Andrew Halpin, and John Murphey, until "Hillsboro" became "Little Ireland." All bought the marginal farmland that the Hills and a few other families had abandoned so recently. With only the plague-impoverished Irish to value it, these lands could be bought very cheaply. By 1860, while there were at least fifteen Hill farms in the rest of Starksboro, there were only two in Hillsboro; by 1870, there was only one. While the first Catholic child born in Starksboro (in 1852) had to be carried by his mother to Burlington to be baptized, this soon changed. Mass began being held in the schoolhouse down the road from what used to be Lemuel Hill's place, and in the spring of 1860, Thomas Casey buried his wife Anora in a plot just south of the schoolhouse a few yards away from a babbling brook that supplied cool water for the creamery he had built. Thomas buried her as if he were back in the Parish of Rooan, County Clare, in a stone crypt capped by a large lintel. Instead of heather there was lilac and live-forever to plant at the grave.

But even for the Irish, Hillsboro was only a landing place. Forces as great as the Famine were working in their adopted nation. The market for wool, which was strong during the Civil War years with the cutoff of southern cotton and the need for immense numbers of uniforms and blankets for Union soldiers, vanished, especially as cheaper western wool moved east via newly constructed rail lines. Holsteins and Jerseys replaced Merinos and Saxonies on Hillsboro's closely cropped hill pastures, but dairying was only a stopgap to the flood of change still coming. The twentieth century kept coming, and with it came farm abandonment.

It all seemed to start with the Civil War. Though far from the battlefield, Hillsboro, South Starksboro, Huntington, Lincoln, and every other Vermont village lost a great deal in the war. One out of every nine Vermonters marched off, and aside from the more than five thousand who were killed, many of the young men, often called away in the middle of haying, started new lives elsewhere. For those who did return, ties to their towns had been loosened by travel, and often they made only a brief stay before heading west to Kansas or Iowa. Not only veterans, but their younger brothers and sisters also left the farms in numbers for more promising opportunities. Those who stayed worried that the state was being "skimmed of the cream" of its population. They pointed to the fact that Vermont was known as the "seed-bed of the nation," citing the statistic that one in a thousand of Vermont's population merited a biography in *Who's Who,* while the other states averaged only .75 per thousand. Taking stock of this situation, certain Vermont leaders asked these questions: "Is the seed-bed of the nation being kept up to former excellence? Will the breeding stock of the future be as virile as it has been? If not, is the seed deteriorating in quality or are Vermonters neglecting to keep the soil of their seed-bed—the physical and social environment of their children—rich, mellow, and weed-free?"

This question was phrased by Henry Farnham Perkins, professor of zoology at the University of Vermont. Son of Vermont's greatest nineteenth-century answerman, George Henry Perkins, who served variously as state naturalist, state geologist, and state entomologist, "Harry" began teaching at UVM in 1902. His special domain was heredity, which he taught until his retirement in 1945. For a biologist, Perkins's interests ran considerably toward social, rather than natural, science. "I look upon the rural community as a biological organism in which the interactions between the organism, its heredity, and its environment show numerous parallels to the kindred processes in plant or animal," said Perkins. Looking down at this organism through the microscope of his science, Perkins saw two competing forces influencing the organism's life history. First, there were "good families"—families that through a succession of generations had improved their towns, intelligent, "socially-minded," responsible, fair and honest families. The Hills were such a family, as of course were Perkins's own—good, Protestant, God-fearing, hard-working.

Then there was the rural community's other life force, the "bad" family. These families, because of inertia, poor judgment, dissipation of energy, or "downright criminal tendencies" acted as a restraint upon progress, nullifying the constructive efforts of their neighbors. These were the "defectives," the "undesirables"—drunks and village idiots and paupers and suicides and prostitutes. They were the genetically unfit who cluttered the Vermont landscape with tarpaper shacks and filled the state's poor farms and mental hospitals. They were the polluted protoplasm that threatened to overwhelm Vermont's pure, original Protestant stock.

Harry Perkins had never been to Hillsboro, or to Jerusalem or Hanksville or Shaker Hill, but as director of the Eugenics Survey of Vermont he knew just where in these villages the tainted protoplasm existed. In his files he had profiles of over 6000 people, and twenty-foot long pedigree charts of fifty-two families. In his office in Burlington he had the data to show which of the branches of Vermont's ancestral tree were diseased and rotten, so that they might be pruned. The tools needed for the work were unassuming—a typewriter, graph paper, colored index cards, maps, and outside, a shiny new '25 Ford. The impressive library that lined the walls of the little office contained the ideological manuals for the use of these tools: shelves of scientific journals like *American Breeder's Magazine,* the *Journal of Heredity,* and *Eugenical News;* pamphlets with titles such as "Race Suicide in the United States," "The Future of America: A Biological Forecast," "The Biology of Superiority," and the "Proceedings of the First Race Betterment Conference"; weighty textbooks such as W. E. D. Stokes's *The Right to Be Well-Born,* and Madison Grant's *The Passing of the Great Race.*

"Eugenics," from the Greek words meaning "well bred," was a term coined by a cousin of Charles Darwin's, Francis Galton, to denote the "science of improving the human race." The "science" of eugenics was an important ally of modern racism—its proponents believed in the existence of racial stereotypes, accepted the myth that certain races (particularly that of northern Europe) possessed a monopoly of desired characteristics, and thought that racial differences were invariably caused by heredity and thus were resistant to modification. It was a pseudo-science born out of genetics, itself only an infant science. The impetus for eugenics in Vermont came partly from a report of the Draft Board in 1918 that suggested that there was an excessive amount of feeblemindedness and other mental defects among the young Vermonters of draft age.

The eugenics "program" consisted of two parts: "positive" eugenics, aimed at giving the public the "facts" of heredity in the hope that "superior" couples would heed the message to have more children; and "negative" eugenics. From the outset, eugenics concentrated on the "negative" approach to its subject, perhaps because early genetic research focused on pathological traits for study. Society's "defectives"—epileptics, criminals, alcoholics, and the insane—were examined with an eye toward eliminating their undesirable traits via marriage restrictions, permanent custody in institutions, and, ultimately, sterilization. The center for the American eugenics movement—for it was indeed more a social crusade than an objective science—was the Eugenics Record Office at Cold Spring Harbor, Long Island.

From that office on September 24, 1925, came a letter from Harry H. Laughlin, Superintendent, to Miss Harriet E. Abbott, the first field worker of the Vermont Eugenics Survey. Laughlin welcomed the Survey to the Eugenics Research Association, and went on to address questions asked by Miss Abbott regarding the practicability of a eugenical sterilization law. Miss Abbott, and later other "field workers," traveled throughout the state to root out the degenerate as well as the "borderline" cases. But when Miss Abbott or her successor Martha Wadman drove the big black Ford up the steep backroads of Lincoln and Starksboro and scores of other towns to visit with families, they didn't talk of sterilization or even of "defectives." They talked smilingly of family history, of genealogy, and of the weather. They wrote polite follow-up letters to farmer's wives: "Thanks for telling me about your family. It was a pleasant as well as profitable call, although I felt a little guilty about coming when you were so busy. This has been a terrible week to move—or to do anything but sit in the shade. I hope you have not had to do all that extra work in this heat." When Mrs. Wadman wrote to relatives who had settled in Wisconsin or Ohio asking for genealogical information, she often had to explain herself: "No, I am not a member of the —— family, but I have had such pleasant visits with so many of them that I feel as if I *were* one of the family." The relations out in the Midwest were delighted that someone was writing a book of their family genealogy, and asked to be sent a copy when it was finished. No, she explained apologetically, there would be no book, not for the time being.

Instead, the genealogical information culled from hundreds of unsuspecting informants from all over the state went into file cards and elaborate pedigree charts that showed eugenicists just where the "bad seed" in these families originated and where it spread. By the end of 1925, the survey had pedigree charts for sixty-two families, including some 4642 people. Of these, 766 were paupers, 380 were "feebleminded," 119 had prison records. There were 73 illegitimate children, 202 sex offenders, and 75 with physical defects such as blindness or paralysis. These figures, Perkins asserted, were "unquestionably much too low . . . the final report will certainly give much higher numbers, especially for illegitimacy and sex offenders." In the front of the top drawer of the main file in the Eugenics Survey office on Church Street in Burlington there was an unlabeled brown binder. The first page was a map of Vermont, many of its towns shaded with colored pencils. The legend on the map explained the colors: Blue towns were those with a significant proportion of the town's population descended from "original stock"; green represented "desirable or progressive" towns; towns colored brown had "declined in some way"; yellow meant that a town had large numbers of "summer people," while grey signified towns with many "outlanders." Following the map were neatly drawn graphs of the population of many Vermont towns, typically showing a steady climb from first settlement in the late 1700s until sometime in the mid to late 1800s, when the lines almost invariably turned downward again. Each graph preceded a brief report about a town, with a list of merchants, doctors, lawyers, and ministers accompanied by characterizations of the town's social, economic, and moral "climate." Each town report contained lists of "degenerate familes," "families who are not very good," and "large families."

Upon arriving in a town, the field workers of the Eugenics Survey—women who usually had some training in social work and who often had received training in eugenic field work at the Eugenics Record Office—would start their research with one talkative informant—the town clerk, a minister, shopkeeper, and soon—and branch out from there. In November of 1929, Martha Wadman took her science out over Hillsboro Mountain to Huntington Village, in her words, "a very attractive place with neat, well-painted houses and well-kept lawns." There she spent an afternoon chatting with the Congregationalist minister Arthur Clarke and his wife. Though they had only been in town for two years, the Clarkes had plenty to say about Huntington and its people. "Most of the people here are all well-to-do," said Mrs. Clarke. "They don't feel that they need to trust in the Lord for anything for they have enough wood to last them two years in their sheds, enough vegetables for the winter in their cellars, and they see no prospect of lacking anything." But not all of Huntington was prosperous. "Hanksville has been the dumping ground for the town, and most of the low-grade families have gathered there. I guess this is because it's more isolated there, and land is cheap. I know there's been a lot of inbreeding down there," concluded the minister's wife.

Were there any particularly "bad" families, Mrs. Wadman wanted to know. The Clarkes started with families they thought "not very good."

There was one family of French Catholics recently come from Quebec. Of the twenty-two children, most were "subnormal," and a few "rather low," according to the Clarkes, who, like most middle-class Americans of the day, were completely comfortable using the language of intelligence testing to describe fellow community members. The Clarkes described another family as "not of the highest caliber but none of them are known to be really low grade. They are more or less spreading over the town."

To get more information about Hanksville's bad seed, Mrs. Wadman needed other informants. Mrs. Clarke suggested a number of neighbors, but cautioned that they be approached carefully. "If they think an outsider is critical of their neighbors, even when they themselves are just as critical, they won't say a word," warned the Clarkes. Dr. Falby of Huntington Village had tended to some of the families in question, but the doctor's wife would be difficult since she was deaf, and the doctor himself might not be cooperative because he was so jealous of any one in town who used another doctor. Reverend Clarke encouraged Mrs. Wadman to interview a woman who had taught at the Hanksville school until the fall of 1928. "She would know all about these low-grade families," assured Reverend Clarke. "If you went to see her on some other errand, and then asked about Hanksville, saying I'd suggested you talk to her, she might help. I think the idea of the Rural Survey would appeal to her. She might be the way to get on better terms with some of the others."

Mrs. Wadman never bothered to talk to any of these people, but she did find out from the Clarkes who the really degenerate families in the area were. "Take ——," said the Clarkes. "He lives next to —— in Hanksville, and has had lots of children. I can't say exactly how many, but the youngest one, a baby, died of starvation about a month ago. Dr. Falby said it was a mere skeleton. All the children are feeble-minded. The two-year old is very fat, and sits on the floor wherever its put and doesn't move. I don't know whether its condition is due to malnutrition or whether it's simply low-grade. From what I hear, the ten-year old used to drop on the floor when he got to school in the morning, he was so tired and hungry. One of the daughters married a psychopath, and they have two children, but her sister luckily has none. The father spends all the money he can get hold of on automobiles. He's crazy about them. He had an accident years back and can't work much now, except for what work he can get at the mill. The family live like cattle, if you ask me."

Reverend Clarke continued with more horror stories. "This fellow's brother just married his second wife. She is a simpleton, and I refused to marry them, knowing her condition. In church when I announce the hymn number, she can remember it long enough to find the hymn only if it's a two-digit number. If it's three digits, forget it." Reverend Clarke then regaled Mrs. Wadman with stories about the most eccentric and feared person in town, an old man with a long white beard and strange manners. No one drank from the spring where this man took his water, for fear they would go crazy too. One of the most frequently heard stories was that when his wife died, the selectmen had to force him to bury her. He said he was too busy with haying. "His children are all subnormal," said Rev. Clarke. "The thirteen-year-old is only in the second grade. Of the older boys, there's one who lives at the Center. His family owns only two chairs—one with a seat and one without. Before him, the family that lived in the house kept chickens upstairs. They have dogs instead of chickens, and after they've starved to death, they throw the carcasses out the window. He does some odd jobs, but every winter it's the neighbors, church people, and storekeepers who help them survive. Another son has a daughter who lives in Starksboro. She didn't get through the first grade, and her husband, a farm-hand, is tongue-tied and not brilliant. I think she was pregnant when she asked me to marry them, but I refused, not because she was pregnant, but because she was low-grade."

Mrs. Clarke knew there had been a lot of intermarriage between a number of these families. "All the ——s in town and over in Starksboro are related in some way, and there's been a number of marriages between cousins. Most all of the branches are fairly normal, but there's one branch that seems to run down. Take ——; he and his family are fine people. Come to think of it though, one of his daughters had a still birth that had a queer flat head. Doctor Falby told her not to have any more children. I guess he felt there'd been too many cousin marriages. ——'s wife was a member of this family—the daughter of second cousins—and she's subnormal. Then there's her half-sister and her mother, who had six illegitimate children all by different men. She worked for Mrs. Falby for a while, but Mrs. Falby couldn't keep her because she was so man-crazy. She got married recently to a terrible creature—I can't remember his name—and they live in the back hills of Starksboro. All their children are extremely low-grade."

Sometimes when Mrs. Wadman went back to her field notes there was some confusion. In her final typewritten report on Huntington, under "Informants," she has listed "—— and ——: it is not quite certain whether or not these were named as informants or as poor families." This sort of confusion was not uncommon. But when it came time for publishing the "data" she and others had collected, clarity and consistency were not necessary. Unsubstantiated anecdotes found their way into the annual reports of the survey as hard data. To a scientifically unsophisticated public, the reports did appear scientific, with their clock-dial pedigree charts and long tables, and since the reports were edited by Professor Perkins, a scientific man who never failed to stress the objective, impartial motivation of the Eugenics Survey, they went largely unchallenged.

There were two key reasons that the Eugenics Survey met with little professional criticism. First, the "science" of genetics was a fledgling science at best, born only two decades before when in 1900 Gregor Mendel's 1866 paper on crossing sweet peas was rediscovered. Before Mendel, there had been no organized eugenics program—the vestigial belief in the inheritance of acquired characteristics and other archaic notions of inheritance stymied the principles of eugenics. But after 1900, Mendel gave the eugenics movement its biological mechanisms and its experimental method. Progress was slow; in 1910, C. D. Darlington estimated that there were perhaps six or eight people in the world who had a thorough grasp of Mendelian genetics. On the other hand, there were scores of zealous social reformers who claimed to understand the science of heredity. Of the more than one hundred people on the Advisory Council of the American Eugenics Society in the mid 1920s, only 10% were trained geneticists. In Vermont, there was not a single geneticist who could pit his scientific authority against the Eugenics Survey.

The second factor that favored Perkins and other eugenicists' claims was the anonymity of their data. The names of the "degenerate" families never made it into the survey's annual reports (unlike the Overseer of the Poor's reports, where each year all those who wanted could read just which of their neighbors had gone "on the town"). The family feuds and gossip of small towns were amalgamated into tidy narratives under fictitious names. The fictitious name as a tool of eugenics research got its greatest boost in 1912 from Henry Goddard, superintendent of the Vineland School for the Feebleminded, with the publication of his book *The Kallikak Family: A Study in the Heredity of Feeblemindedness.* Goddard had forged the pseudonymn for the family he studied from the Greek *kallos* (meaning "beauty") and *kakos* (meaning "bad"). The intended implication was that the "good" hereditary strain had been tainted by the "bad." "Kallikak" became synonymous with "feebleminded." Perkins's pseudonymns, however, did not ring with classical overtones. The second annual report contained the story of the "Doolittle" family, many of whose members were a "moral, social, and economic drag on town and state from the very outset." The following year the report dropped the embarrasingly obvious "Doolittle" epithet and instead considered the "Furman" family of "Garfield." "Matthew Furman" came from southern New Hampshire in 1803, buying about 300 acres on the hills above the village, which he cleared, along with building the town's first grist mill. Matthew had eight children, and in 1929 there were 372 direct descendants.

The Furman narrative painted a familiar picture. Members of the Society of Friends, the family showed all the Quaker virtues—industry, thrift, and simple living.

> As an illustration of their simple living a story is told of one of the family, who, being taken to an hotel to dine when visiting relatives in Boston, quietly ordered a bowl of bread and milk. The children were often sent to Quaker school in Ohio to finish their education. Three members of the direct line and one mate have been Quaker ministers. . . . The family as a whole is law abiding, highly respected, and its members are usually successful in whatever they undertake.

Like the Hills and many other Vermont families, the "Furmans" had a strong feeling of family solidarity.

> The family antiques are highly prized as mementos of the past. Speaking of her recent marriage of her brother to a member of another branch of the family who had inherited some of the Furman furniture, one woman added that they were especially glad because now these antiques would come back to the family again. The various branches of the family are always in touch with one another, and visiting and annual reunions show their pride in their blood. Only recently one member, on his wedding trip, made a tour of the country visiting relatives. The family reunions have been held for the last fifty years and are usually attended by as many as one hundred people.

But there was a dark side to the Furman family.

> Insanity is the family's only defect. The six known cases—including two suicides—and two more suicides not known to be insane, are all in one branch. Evidences of insanity appear elsewhere—one often hears of individuals who

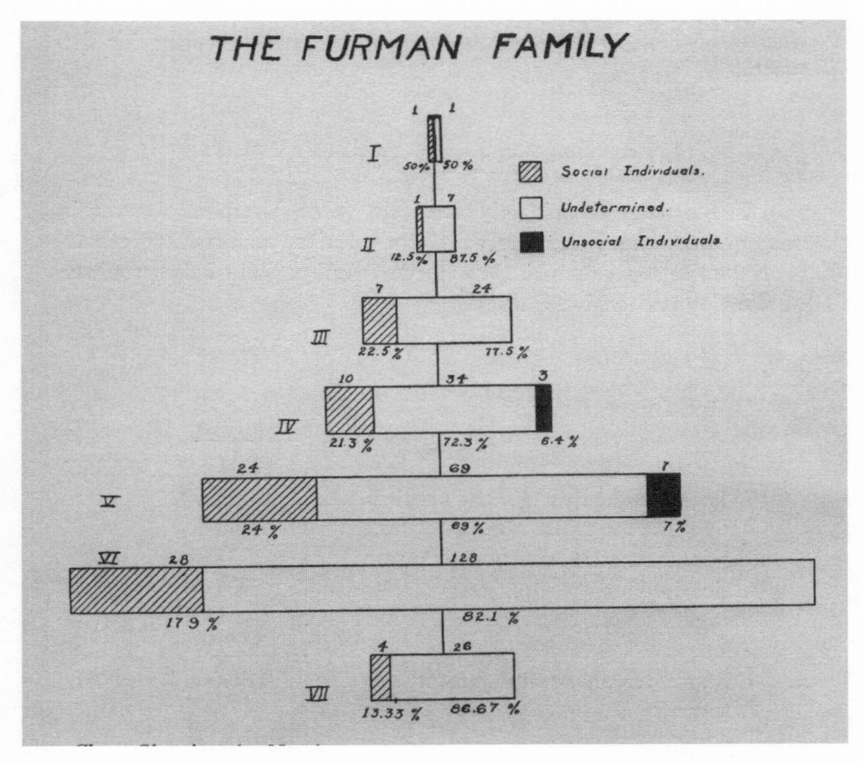

Chart showing the "number and percent of Social and Unsocial Individuals in the Direct Line of the Furman Family, by Generations." According to the chart legend, "Social" individuals are "desirable citizens—law-abiding, self-supporting, and doing some useful work"; "Unsocial" individuals consist of the "insane and suicides"; the "Undetermined" category comprises "those who, while not showing either of the defects, do not seem to show any socially desirable tendencies, and those about whom too little is known to make any judgment possible." From *Third Annual Report of the Eugenics Survey of Vermont, 1929.*

> were queer, had nervous breakdowns, or were sick for long periods of time. One such is the case of a girl of 19 who is said to have committed suicide because her parents were so strict and kept her shut up. Other informants say, however, that she had been ill for a year and a half. These various statements suggest the possibility that she was insane.

These "various statements" suggest also the invalidity of the entire eugenics enterprise. Hearsay, innuendo, grudges, and plain lies were woven together and presented as safely anonymous fact. The "Furman" history was built from informants in the same manner as Mrs. Wadman's Huntington report. The Furman text is a verbatim rendering of information gathered about a well-known, and well-loved, family of the Bristol/Lincoln/Starksboro area. Rowland Robinson knew this family's name, which he borrowed for a number of his most beloved fictional characters. As certainly as Robinson bestowed certain familiar qualities on these characters, Perkins bestowed upon the Furmans the traits he desired to portray. He did this again in the 1930 Annual Report, via the "Burrs," another "Garfield" family. The "Burrs" were actually Bristol selectman and state representative Truman Crane Varney's family, whom Perkins cast in the role of a "good" family who, although some members had moved away, still contributed to the vitality of their town.

Perkins's myth making rang true for most of his contemporaries. In 1929, none of this sounded sinister to the people of Starksboro; their senator, William P. Dillingham, had headed the effort to restrict immigration into the United States, to keep the alien from their midst. After Dillingham, it was their senator, Frank L. Greene, who made long and much-applauded speeches against paternalism in government, and who championed the "ancient Green Mountain Boys' tradition" as the one that should guide not only Vermont, but the world. The world was changing—even Starksboro's little hill communities felt it. Electric lights were coming to towns once lit only by oil lamps and candles, and automobiles raced along past horses on the road from Bristol to Burlington. People pulled in the reins, via immigration restriction, prohibition, even via eugenical sterilization.

From the outset, the passage of a sterilization law in Vermont was the primary aim of the Eugenics Survey. A week after the survey set up its office, a letter came from Harry H. Laughlin, superintendent of the Eugenics Record Office at Cold Spring Harbor, New York, to Miss Harriet Abbott, the survey's first "field worker." Replying to Miss Abbott's inquiry regarding the practicability of a eugenical sterilization law in Vermont, Laughlin replied:

> Sterilization will simply be one of the several remedies which the state ultimately will use in applying eugenics to population control. At the present state of biological and psychiatric knowledge, a great deal is known about human heredity—enough to make eugenical sterilization a safe policy, provided the standards for sterilization apply only to the most patently degenerate individuals. . . . In the future, the standards can be shifted to include those individuals who now constitute situations described as "border-line."

A month later the Advisory Committee of the Eugenics Survey endorsed the passage of a eugenical sterilization law in Vermont. Along with Perkins, Miss Abbott, and the superintendents of the various state institutions, the

Advisory Committee included Guy Bailey, president of the University of Vermont, A. R. Gifford, president of the Vermont Children's Aid Society, and Clarence Dempsey, the Commissioner of Education. In 1927, the Vermont Senate passed a bill for a sterilization law, but it was rejected by the House. By 1929, after the Great Depression had leveled rich and poor, "ill-bred" and "well-bred" alike, eugenics began to lose ground in America, but in Vermont it was still gathering steam. Vermont was far from Wall Street; though farms were hard hit by the financial crisis, there were no wealthy magnates jumping out tenth-story windows. Emigration and the supposed consequent decline of the hill farm was considered the main problem in Vermont. Senate Bill 64, "An act for human betterment by voluntary sterilization," introduced in February 1931, could only help deal with the social ills that stemmed from this "skimming of the cream" of Vermont's population. The Senate passed the bill 22 to 8, the House 140 to 75. The representative from Starksboro voted for the bill, as did Truman Crane Varney of Bristol. His cousin George Varney had spent many pleasant hours talking genealogy with Miss Wadman. On April Fool's Day, 1931, Governor John Weeks signed Act 174—"An act for human betterment by voluntary sterilization"—into law.

Today the area that was Hillsboro makes up the Lewis Creek Wildlife Management Area, and the farms that once sustained the Hills and the Irish have reverted to forest, sustaining again their original inhabitants—deer and ruffed grouse and red fox. Cellar holes and overgrown orchards speak of former days, as do the cemeteries. They contain the bones of Hillsboro's entire brief but varied human history. In the Butler Cemetery, at the top of the steep ravine on Hillsboro's southern flank, amid a sea of periwinkle, lie Nortons and Bennetts and Jameses and Hallocks. Up the road is Thomas Casey's stone crypt, surrounded by Conways and Hannans. On the northern margin of the quieted community, the little grassy knob whose headstones all face west is called the Hallock Cemetery; along with Hallocks there are Hills, Browns, Sweets, Shattucks, and others. In the "center" of Hillsboro, if ever there was such a thing, is the Hillsboro Cemetery, surrounded by a crumbling but still well-defined stone wall punctuated by open-grown sugar maples. An old apple tree stands at the entrance, spilling fruit for mice and deer. The seeds lie naked and unrooted on the ground. In the foreground is a granite headstone in the form of a severed tree, its branches all cut back to the trunk. Carved on the trunk is this memorial:

TRUE W. LAYN
DIED
SEPT. 11, 1902
AGE 79 YRS., 10 MOS.
HIS WIFE
ABIGAIL HILL
AGE 63 YRS., 2 MOS.

Their children
William Guy (age 5) died March 27, 1857
Mary E. (age 3) died June 13, 1862
Guy (age 1) died April 13, 1849
Abbie (age 26) January 10, 1888

None of True and Abigail's children outlived them; they were not alone—of the 28 people buried in the cemetery, only eleven lived past age thirty, and ten did not live past age ten. The Layns' family tree was severed by disease, by accident, and by fate. Its branchings were like the sugar maple's and Lewis Creek's—random and never wholly permanent. Each branch sought life as surely as the maple's new shoots sought sun, or as persistently as the headwater streams eroded headward into Hillsboro's hills. The Layn family's branches were cut by chance.

In Starksboro and throughout the Lewis Creek watershed there are other cemeteries that hold the bones of those whose branching was severed by the state, empowered by Act 174. In the first decades of the twentieth century, families of French-Canadian and Abenaki descent who had for generations made the lower reaches of Lewis Creek their seasonal home transformed their canal barges into houseboats after the demise of Lake Champlain's lumber boom. Rowland Robinson's beloved fictional compatriot of Sam Lovel, Antoine Bassette, was no doubt a composite portrait of some of the men of these families. The Eugenics Survey of Vermont painted its own composite portrait: These people were "Pirates," who lived in "the utmost squalor and destitution," and were the "terror of people who own boats or other property in the neighborhood because of their thieving habits." The Eugenics Survey pedigree chart concluded that "their most characteristic habit is to hunt and fish out of season and to cleverly vanish when in danger of being caught, then appear in a new place." Harry Perkins knew these "pirate" families well, as they often moored their "shanty boats" near the Burlington waterfront. Around Burlington, the most famous "pirate" was

Sylvester Ploof. Friends knew him as a man of the water, who sailed up and down the lake transporting sand, gravel, lumber—anything at all—from one port to another. In 1890, he had married fourteen-year-old Ellen Shappy, and for sixteen years they plied the lake together, Mrs. Ploof giving birth to nine children, whom she proudly declared could "swim like fishes." They settled down occasionally in Burlington, where Ellen found work in a grocery store while Sylvester worked as a teamster.

In early November of 1904, Sylvester Ploof set out on his houseboat from the bay just north of Hawkins Bay, to shuttle another "pirate" family across Lake Champlain to Willsboro, New York. Shortly after they cleared the headlands of Converse Bay, a tremendous storm blew over the Lake. Sylvester immediately steered them for a nearby island—Garden Island, almost a twin to Gardiner's Island—to wait out the storm. After they had moored, the caretaker of the island, who had been given instructions by the owner, New York City hardware and barbed wire magnate Henry W. Putnam, not to let boats tie up on his dock, came out and ordered Ploof to leave. When Ploof refused, the caretaker set the houseboat adrift. A short time later, as Ploof and his passengers watched helplessly after swimming to shore, the houseboat was smashed to pieces on the rocks. Sylvester had died by the time Perkins's Eugenics Survey began, but his widow still lived in a houseboat inside the Burlington breakwater, and several other "pirate" families lived not far away. Sylvester, Ellen, and their children all had cards in the files of the Eugenics Survey. The class prejudice of late-nineteenth-century Vermont that sent Sylvester Ploof's home and family smashing into the rocks ringing Converse Bay found new power through pedigree charts, 3 by 5 cards, and the sterilization law.

The Tahmonts, Taxuses, and other Abenaki families whom Rowland Robinson met along Lewis Creek and Little Otter on their trips down from Odanak always made a point of visiting with their local relations, Abenaki who had never left Vermont, but stayed close to the rivers and Lake of their ancestors. To Harry Perkins and the Eugenics Survey, the Abenaki families of the Champlain Valley were "Gypsies": "This whole family numbering well over 150 individuals, retains its ancestors' roving or Gypsy tendency. They are horse traders, fortune tellers, and basket makers." One of the Abenaki families who had long made their home in the Lewis Creek and surrounding watersheds became Pedigree #2 of the Eugenics Survey records. ESV fieldworker Harriet Abbott spent many hours interviewing an Abenaki man who lived in Salisbury, in a two-room shack on the side of the railroad right-of-way, about a half-mile from Otter Creek and less than a mile from the Cornwall cedar swamp. It was the perfect place for a "gypsy," out of sight of townfolk and close to the land where he could hunt, fish and trap unmolested. With his wife, —— had lived all over this part of the Champlain Valley—Vergennes, Panton, New Haven, and now Salisbury, where he worked as a section hand on the railroad. Abbott arrived there about dusk, and in a little while along came ——, driving his old horse very slowly toward home. This was just like ——, Abbott later reported, having heard from people in the village that he was always buying derelict horses and trading or selling them. —— told Abbott that it was some family's pet, and they had asked him to kill it because it was so old; finding the old horse to his liking, it had become his trusted mount.

——'s sleigh was as derelict as his horse, as Abbott discovered when he drove her to the station to catch a train back to Burlington. "When Agent H. E. A. had seated herself in the various comfortables, that had been brought from the house, she promptly fell through, revealing the fact that there was no cover to the top of the seat but that the holes were filled in with comfortables, etc." —— and his wife had made every effort to make the lady from Burlington feel at home. Abbott noted that —— had given her more information about his relatives than any other individual. Without him, she could never have begun the "gypsy" pedigree chart. Their hospitality was rewarded with these words from Harriet Abbott in her report: "The people in the village rather laugh at ——, take it for granted that he is part negro, and say that he can not help but trade horses, etc. They, however, have no fault to find with his honesty—he apparently has settled down very well for a ——." It seemed obvious to Abbott that by marrying his light-haired, blue-eyed wife ("who is rather more intelligent than he") —— was trying to "take a step upward." Though —— was clearly "subnormal," according to the generous estimation of Abbott he was "far above" his father in intelligence, and above the "definitely feebleminded class."

Maude Catchapaw's clan was one of the other branches eyed suspiciously by the Eugenics Survey. The Catchapaws' friend and neighbor in Hinesburg, Representative Lyle Baldwin, likely never realized when he voted for the sterilization bill that it would be aimed at his neighbors' kin.

The apple tree by the Hillsboro Cemetery gateway bears dozens of pruned branches. Each was pruned to beautify and strengthen the tree, but what of the possibilities of each pruned branch? What unique and wonderful flower and fruit might they have borne? Lewis Creek too bears the scars of constant pruning, in its oxbow sloughs cut off from the main channel, and in each of its many cut banks, where it relentlessly does its erosional work upon the land. For tens of thousands of years, the soil that it has cut from its margins has been transported downstream, deposited on point bars or carried all the way past the Myers Landing and Jigwallick, out into Hawkins Bay, to settle out near the shores of Gardiner's Island. Like the fossil-laden sediments on an ancient Cambrian shore once did, they await their future chronicler.

"Pruning" is an act of controlling nature, a radical reshaping of the forces that give the physical world its form. The eugenicists in Vermont and the rest of America seem to us now to have committed a violent act of inhumanity upon their fellow citizens, but in their day they would have accepted the "pruning" metaphor as an acknowledgement of their careful stewardship of the social and biological landscape. It is difficult for us to imagine just how "progressive" eugenics was considered by educated Vermonters of the 1920s and 1930s. Harry Perkins and his fellow eugenicists would no doubt be disappointed in subsequent generations' failure to carry on this stewardship; indeed, in the 1950s, Perkins lamented that eugenics had lost steam, and maintained that it was one of the most important social movements in Vermont history. Rowland Robinson would have offered another assessment; keenly aware of the limitations of science, and the limitlessness of human folly, Robinson would likely have railed against eugenics with the sort of passion his father voiced against slavery. (And yet, Robinson's own son Rowland T. was a Eugenics Survey of Vermont "informant.")

It is hard to recover the voices of both those subject to the eugenicists' social programme, as well as those who were opposed to it. But perhaps the most clear-headed assessment came from a man born and raised in the Lewis Creek watershed, E. F. Johnstone of Bristol. In 1931, he sent his composition, "Authority to Mutilate," to the *Rutland Herald:*

> Hide, ye lunatics and fools,
> They are ready with their tools,
> Only waiting the approval of the state.
> When they think it safe and fit
> They will make a man of it
> If they get authority to mutilate.

They will hunt them in the hills,
in the factories and mills,
  They will chase them through the
  counties sure as fate,
If they find a man alone
They will swat him with a bone
  In their eagerness and haste to mutilate.

They will sterilize the lads
With their uncles and their dads—
  Any person not exactly up-to-date
They will do it for a price
And I offer my advice—
  Give them no authority to mutilate.

It will lead to great abuse
If they turn the doctors loose
  With their tomahawks and lances
  up-to-date.
There'll be eunuchs short and tall
And the count will not be small
  If you grant to them the right to mutilate.

When a doctor wants a boat
On the broad highways to float
  He will find a place where sapheads congregate
He will chase them to a shed
And at fifty bucks a head
  He will freeze his conscience out and mutilate.

Let them sterilize the cats,
Evolutionists and rats,
  And the gypsy moths that widely circulate.
Working at their own expense,
None will ever take offense;
  For you may be sure they'll never operate.

People all over the country have found the story of eugenics in Vermont, superbly told by historian Nancy Gallagher in her 1999 book, *Breeding Better Vermonters: The Eugenics Project in the Green Mountain State,* to be shocking. Yet, Vermont was the 31st state to pass a law permitting eugenical sterilization; neighboring New Hampshire passed its first "voluntary" sterilization statute in 1917, revised it in 1929 to encompass a wider spectrum of the New Hampshire population and permit compulsory sterilization, and is on record as having sterilized more of its citizens than Vermont. California sterilized tens of thousands of its residents. Americans, cherishing an image of Vermont as a uniquely egalitarian place, have been particularly outraged to discover here in their imagined Arcadian refuge a dark episode of discrimination, prejudice, and coercive laws in violation of basic human rights. All landscapes have emotional qualities bestowed upon them by the human imagination, and Americans for over a century have given uniformly positive ones to Vermont. Where today we might regard the lifestyles of these families—living close to the land as they did—as admirable, to the eugenicists—many of whom were ardent conservationists—such lifestyles were seen as confirmation that those who led them were degenerate and depraved. Eugenics was in many ways a program of landscape alteration whose aim was to replace Vermont's tarpaper shacks, canal boats, and canvas tents with well-kept homes of "summer people."

Standing on the red rock bluffs of Mount Philo, looking southwest to Lewis Creek's mouth and Gardiner's Island, one can also see Garden Island and Converse Bay, where Sylvester Ploof's canal boat came smashing onto shore. One can make out cedar swamps and other marginal places that occasionally served as havens for the "pirates" and the "gypsies." From the Starksboro and Huntington ridges that hold the headwaters of Lewis Creek, one is but a brief walk from the cellar holes of families visited in the 1920s by eugenics fieldworkers. Looking out from any of these Lewis Creek prospects today, the adages of Henry Perkins and his fellow eugenicists'—"Blood has told"; "Running water purifies itself"—ring in one's ears. But the metaphors ring hollow and flat, given our historical perspective. The penetration of eugenical thinking into the "blood streams" of Lewis Creek and the rest of Vermont may be the most critical episode in the region's history of "finding" and "losing." Believing that they had found bad blood streams, the eugenicists sought to lose them from the stream of Time, to force the river of life down one scientifically engineered, rip-rapped channel. Having "found" this troubling history, it is our responsibility to learn from it, in this era when we stand so powerfully as "makers" rather than "finders" of Nature.

Conclusion

I would fain say something, not so much concerning the Chinese or Sandwich Islanders as you who read these pages, who are said to live in New England.
—Henry David Thoreau

N THE 1920s and 1930s, rural poverty was perceived by eugenicists as a threat to the harmony of Vermont's landscape; today, the surge of residential development caused by prosperity is the main concern of many Lewis Creek watershed residents. While eugenicists were determined to selectively rid the landscape of its vestiges of a particular rural past—"degeneration" after all, was a term that suggested a return to a prior condition—contemporary environmental activists are equally determined to selectively preserve and restore the landscape to a condition where the "degenerate" gypsies and pirates and hill folk would perhaps have been most comfortable. In the last decade, under the auspices of the Lewis Creek Association, hundreds of people have kept a vigil upon the Creek that would do Rowland Robinson proud. The association conducts a vigorous volunteer program of water quality monitoring, streambank stabilization and restoration, wildlife censusing, outdoor education, and land protection activities. Many of the people who joined the association were first motivated to do so by a pronounced sense of loss, after some feature of the Lewis Creek landscape—a woodlot, a working farm, a wetland—was suddenly threatened.

A founding principle of the Lewis Creek Association is that a watershed approach is essential to effective environmental planning. By bringing together citizens from seven towns in two counties, it enlarges the sense of both the biotic and social communities necessary for sustainability. While soil conservation districts, environmental planning districts, river-basin commissions, and other multimunicipal organizations have existed for many decades, these bodies have generated almost exclusively professional allegiance. They have never become vehicles of widespread citizen participation and activism. But the members of the Lewis Creek Association identify strongly with their watershed. At the group's recent annual meeting, held at Rokeby, the Rowland Robinson homestead, a large map of the watershed was displayed, and all were invited to mark the place on the map where they lived. Each person who did this claimed the Lewis Creek watershed as *home*. Even if they had not canoed its navigable length, traveled every tributary brook, or made a pilgrimage to its headwaters, they carried within themselves a distinct sense of the watershed as a natural unit worthy of their attention and celebration.

This emerging allegiance to the watershed as home is itself a "watershed," a defining moment in the history of regional consciousness. Cyrus Pringle, though he knew intimately the warp and woof of the Lewis Creek watershed's green fabric, never conceived of himself as a "watershed" citizen. Rowland Robinson, as fierce as his devotion was to Sungahneetook, thought of himself as a citizen first of Ferrisburg, then of Vermont. John Bulkley Perry employed geomorphological concepts constantly as he ranged through earth history, but he would have found a watershed congregation an odd one indeed. But all three men, and the many residents of the Lewis Creek watershed whose stories I have left untold, are members of a watershed congregation, a congregation that, like the watershed concept itself, we create with our minds.

That the landscapes around us are built up as much from our inner processes as they are from outer processes is an increasingly common, though still contested, credo. For Rowland Robinson, this notion had purchase only in the sense that he felt that his perception of the natural world was wholly active, that he needed to bring acute qualities of attention to the land upon which he loved to ramble. His sketches, paintings, journals, and writings were his perceptual workshop. Cyrus Pringle's workshop was his herbarium, where all his plant perceptions were distilled into single sheets of each species, with the dried specimens arranged to be as faithful as possible to the living plant. Behind each herbarium sheet there lay an array of pictures of the long tramps he had taken to find his quarry. An introspective soul, Pringle was still largely oblivious to his own epistemology. The modern science of systematics speaks of species as idealized entities, "representations" of a human-created system of order. For Pringle and his peers, species were altogether real, and his plant hunting was part of an enormous collective quest to assemble a true and faithful picture of Nature. "Belief" in the late nineteenth century belonged to the realm of the Spirit, not of Nature. John Bulkley Perry's efforts to bridge the two domains suggests just how strictly divided they were then felt to be.

Today's explorers of Lewis Creek, and of all of New England's rivers, are more like Rowland Robinson—skeptical, trusting Nature itself as heaven enough—than Pringle and Perry, who, despite their fidelity to natural science, never faltered in their belief in a higher world behind the veil of Nature. Here perhaps is another watershed at which we stand: At the moment when Nature itself is open to question as a realm of transcendental truth, we find ourselves more than ever in need of it as a bulwark against the "virtuality" of the modern world. The place-making activities such as the ones that Lewis Creek Association members engage in are essential because they take the place of the old economies—farming, forestry, hunting—that once bound most of the watershed's denizens to the land. Pringle and Robinson were farmers first, and they had to struggle when they attempted to replace that economy with plant hunting or writing and sketching. Perry was but once removed from toil upon the land, in that the majority of his congregation were still farmers. This trio of explorers—each of them in some way straddling the line between amateur and professional naturalist—were lucky to have lived in an era when the amateur could still make immense contributions to natural science. Our discoveries—of a new location of a rare plant, an otter's den, or a Paleozoic fossil—are much more likely to be recreational sources of personal satisfaction than practical contributions to scientific knowledge. But our finds can also now be deeply *re-creational* as well, for each act of physical discovery more than ever brings to us psychic and spiritual renewal.

It may be that the very act of reflecting on the idiosyncratic history of natural scientific discovery within the Lewis Creek watershed is also a defining moment. Along the banks of Lewis Creek and every other river in New England, and, indeed, in North America and all across the world, innumerable incidents of discovery occur every day, both by professional naturalists and amateurs. The thrill of finding—whether the object of one's search or some serendipitous surprise—unites us with Pringle, Robinson, and Perry, but we just as surely share with them the tragedies of losing. Sometimes finding, sometimes losing, onward we run like the river to the sea.

References

Introduction

Hill, Ralph N. 1949. *The Winooski: Heartway of Vermont.* Rinehart, New York.
Ritter, Dale F. 1978. *Process Geomorphology*. William C. Brown, Dubuque, Iowa.
Strahler, Arthur N. 1975. *Physical Geography.* John Wiley and Sons, New York.
Thoreau, Henry David. 1849. *A Week on the Concord and Merrimack Rivers.* G. P. Putnam, New York.
Warshall, Peter. 1976. Streaming wisdom. *CoEvolution Quarterly* 12: 4–10.

Chapter 1. Footfall

Carruth, Hayden. 1973. Rowland E. Robinson: Vermont's neglected genius. *Vermont History* 41(4):181–197.
Emmons, Ebenezer. 1842. *Geology of New York, Part III, Survey of the Second Geological District.* White and Visscher, New York.
Graffagnino, Kevin. 1984. *Atlas of Lake Champlain 1779–1780, by William Chambers, R.N.* Vermont Heritage Press, Bennington, Vermont.
Hitchcock, Edward, Edward Hitchcock, Jr., Albert D. Hager, Charles H. Hitchcock. 1861. *Report on the Geology of Vermont—Descriptive, Theoretical, Economical, and Scenographical.* 2 volumes. Claremont Manufacturing Co., Claremont, New Hampshire.
Robinson, R. E. 1901. *Danvis Folks.* Houghton Mifflin, Boston.
———. 1889. *Sam Lovel's Camps.* Forest and Stream, New York.
———. 1934 (1897). *Uncle Lisha's Outing and Along Three Rivers.* Charles E. Tuttle Co., Rutland, Vermont.
———. 1898. *Uncle Lisha's Shop: Life in a Corner of Yankeeland.* Forest and Stream, New York.
Vermont Acts, 1874–1896.
Zika, Peter F. 1986. The distribution of *Quercus muehlenbergii* in Vermont. *Vermont Botanical and Bird Clubs Bulletin* 20:15–19.

MANUSCRIPTS

Letter from Cyrus Pringle to George Engelmann. 12/25/1878. Archives, Missouri Botanical Garden.
Letters from George Englemann to Cyrus Pringle. 12/6/1878, 2/20/1879. 11/30/1879. Archives, University of Vermont.

Chapter 2. Liminal Places and People

Day, Gordon M. 1964. A St. Francis Abenaki vocabulary. *International Journal of American Linguistics* 30(4):371–392.

———. 1978a. Ethnology in the works of Rowland E. Robinson. Pp. 36–39 in *Papers of the Ninth Algonquian Conference,* William Cowan, ed. Carleton University, Ottawa.
———. 1978b. Western Abenaki. Pp. 148–159 in *Handbook of North American Indians: Northeast,* B. G. Trigger, ed. Smithsonian Institution Press, Washington, D.C.
Haviland, W. A. and M. W. Power. 1981. *The Original Vermonters: Native Inhabitants, Past and Present.* University Press of New England, Hanover, New Hampshire.
Huden, John. 1955. Indian place names in Vermont. *Vermont History* 23(3):191–203.
Laughlin, S. B. and D. P. Kibbe. 1985. *The Atlas of Breeding Birds of Vermont.* University Press of New England, Hanover, New Hampshire.
Robinson, Rowland E. 1889. *Sam Lovel's Camps.* Forest and Stream, New York.
———. 1896. *In New England Fields and Woods.* Houghton Mifflin, Boston.
———. 1897. *Uncle Lisha's Outing.* Houghton Mifflin, Boston.
Spargo, John. 1936. Foreword. Pp. 5–8 in Rowland E. Robinson, *Sam Lovel's Boy with Forest and Stream Fables.* Charles E. Tuttle, Rutland, Vermont.
Thompson, Zadock. 1972. *Natural History of Vermont.* Charles E. Tuttle, Rutland, Vermont.

MANUSCRIPTS
Diary entry of Rowland Robinson. August 27, 1870. Robinson Papers, Rokeby Museum, Ferrisburg, Vermont.
Letters from Manly Hardy to Rowland E. Robinson. 10/12/1895, 3/20/1895. Robinson papers, Rokeby Museum, Ferrisburg, Vermont.

Chapter 3. Up Lewis Creek

Day, G. R. 1981. Abenaki place-names in the Champlain Valley. *International Journal of American Linguistics* 47(2):143–171.
———. 1978. Ethnology in the works of Rowland E. Robinson. Pp. 36–39. In *Papers of the Ninth Algonquian Conference,* William Cowan, ed. Carleton University, Ottawa.
Doughty, Robin W. 1975. *Feather Fashions and Bird Preservation: A Study in Nature Protection.* University of California Press, Berkeley.
Eggleston, W. W. 1912. Reminiscences of Cyrus G. Pringle. *Vermont Botanical Club Bulletin* 7:8–11.
Forest and Stream. Issues for March 15, 1875, January 27, 1876, August 14, 1879, and September 25, 1879.
Perkins, Mary R. 1936. Biographical sketch of Rowland Evans Robinson. In *Out of Bondage* (Centennial Edition). Charles E. Tuttle, Rutland, Vermont.
Pringle, C. G. 1897. Reminiscences of botanical rambles in Vermont. *Burlington Free Press,* February 9, 1897.
Robinson, R. E. 1878. Fox hunting in New England. *Scribners Monthly.*
———. 1934. *Uncle Lisha's Outing and Along Three Rivers.* Charles E. Tuttle, Rutland, Vermont.
———. 1895. A voyage in the dark. *Atlantic Monthly.*
Van Oosten, John. 1930. Resumé of the history of the Atlantic salmon in the United States with special reference to its re-establishment in Lake Champlain.
Vermont Fish and Game League Annual Report for 1901.

MANUSCRIPTS
Diary entries of Rowland Evans Robinson. 6/29/1876, 7/28/1880. Robinson Papers, Rokeby Museum, Ferrisburg, Vermont.
"Indian Records" notebook entries. 10/26/1858, 10/1859. Robinson Papers, Rokeby Museum, Ferrisburg, Vermont.
Letter from C. E. Faxon to R. E. Robinson. 8/17/1892. Robinson Papers, Rokeby Museum, Ferrisburg, Vermont.

Letter from G. W. Sears to R. E. Robinson. 4/13/1888. Robinson Papers, Rokeby Museum, Ferrisburg, Vermont.
Letter from J. W. Titcomb to R. E. Robinson. 12/30/1892. Robinson Papers, Rokeby Museum, Ferrisburg, Vermont.
Letters from M. C. Edmunds to R. E. Robinson. 3/1/1875, 5/3/1875, 3/17/1876, 5/8/1876. Robinson Papers, Rokeby Museum, Ferrisburg, Vermont.
Letter from Rowland Evans Robinson to M. C. Edmunds. 4/29/1876. Robinson Papers, Rokeby Museum, Ferrisburg, Vermont.
Letter from W. R. Peake to R. E. Robinson. 11/25/1896. Robinson Papers, Rokeby Museum, Ferrisburg, Vermont.

Chapter 4. Wild Apples

Anderson, Katharine. 1981. *Volunteer Apple Trees in Vermont.* Master's thesis, Department of Geography, University of Vermont.
Fink, Steven. 1986. The language of prophecy: Thoreau's "Wild Apples." *New England Quarterly* 59(2):212–230.
Hedrick, U. P. 1950. *A History of Horticulture in America to 1860.* Oxford University Press, New York.
Pringle, C. G. 1871. The old vs. the new. *Country Gentleman,* May 4, 1871.
———. 1874. Hybrid and cross-bred apples. *Country Gentleman,* January 1, 1874.
Richardson, R. D., Jr. 1986. *Henry Thoreau: A Life of the Mind.* University of California Press, Berkeley.
Robinson, Rowland. 1885. A.D. 1950. *Forest and Stream,* September 17, 1885:142.
———. 1934. *Uncle Lisha's Outing and Along Three Rivers.* Tuttle Company, Rutland, Vermont.
———. 1937. *In New England Fields and Woods.* Tuttle Company, Rutland, Vermont.
Thoreau, H. D. 1893. *Excursions* (volume 9 of *The Writings of Henry David Thoreau,* Riverside Edition). Houghton Mifflin, Cambridge, Massachusetts.
Waugh, F. A. 1895. Hardy apples for cold climates. *Vermont Agricultural Experiment Station Bulletin 61.* Burlington, Vermont.
———. 1900. Apples of the Fameuse type. *Vermont Agricultural Experiment Station Bulletin 83.* Burlington, Vermont.

MANUSCRIPTS
Diary entry, Rowland Robinson. April 23, 1878. Robinson Papers, Rokeby Museum, Ferrisburg, Vermont.
Diary of Joseph Rogers. 1870–1872. Rokeby Museum, Ferrisburg, Vermont.

Chapter 5. . . . and Weeds

Burns, G. P. 1897. Orange hawkweed or "paint-brush." *Vermont Agricultural Experiment Station Bulletin 56.* Burlington, Vermont.
Pringle, C. G. 1872. The weeds of Vermont. *Annual Report of the State Board of Agriculture.* Montpelier, Vermont.
Zika, P. F. 1986. A botanical look at weeds. Unpublished manuscript.
Zika, P. F., R. J. Stern, and H.E. Ahles. 1983. Contributions to the flora of the Lake Champlain Valley, New York and Vermont. *Bulletin of the Torrey Botanical Club* 110(3):366–369.

Chapter 6. Star in a Stoneboat

Baldwin, Henry I. 1979. The distribution of *Pinus banksiana* Lamb. in New England and New York. *Rhodora* 81:549–565.

Beers, F. W. 1871. *Atlas of Addison County, Vermont.* F. W. Beers, A. D. Ellis, and G. G. Soule, New York.

Burlington Free Press, March 28, 1877; March 29, 1877; April 13, 1877.

Fernald, M. L. 1919. Lithological factors limiting the ranges of *Pinus banksiana* and *Thuja occidentalis. Rhodora* 21:41–67.

Frost, Robert. 1921. A Star in a Stoneboat. *Yale Review* 10 (January):259–261.

Palmgren, Alvar. 1929. Chance as an element in plant geography. In *Proceedings of the International Congress of Plant Sciences,* B. M. Duggar, ed. George Banta, Menasha, Wisconsin.

Robinson, Rowland E. 1887. *Uncle Lisha's Shop.* Forest and Stream. New York.

———. 1905. *Out of Bondage and Other Stories.* Houghton Mifflin, Cambridge, Massachusetts.

Sears, J. H. 1881. Notes on the forest trees of Essex, Clinton, and Franklin counties, New York. *Bulletin of the Essex Institute* 13:174–178.

MANUSCRIPTS

Diary, Rowland E. Robinson. 1876. Rokeby Museum, Ferrisburg, Vermont.

RER's account of discovery of *Pinus banksiana* in Vermont. Attached to herbarium specimen of *Pinus banksiana* collected by C. E. Faxon, August 15, 1880. Gray Herbarium, Harvard University, Cambridge, Massachusetts.

Letter from C. E. Faxon to R. E. Robinson. May 13, 1882. Rokeby Museum, Ferrisburg, Vermont.

Letter from John R. McCreary to the author. November 6, 1986.

Journal of Anna Robinson. June 6, 1902. Rokeby Museum, Ferrisburg, Vermont.

Chapter 7. Eden Lost

Brainerd, E. 1911. Cyrus Guernsey Pringle. *Rhodora* 13:225–232.

Cadbury, H. J., ed. 1962. *The Civil War Diary of Cyrus Pringle.* Pendle Hill Pamphlet 122, Lebanon, Pennsylvania.

Darwin, Charles. 1868. *Variation of Animals and Plants Under Domestication.* 2 Volumes. J. Murray, London.

Davis, H. B. 1936. *Life and Work of Cyrus Guernsey Pringle.* Free Press Printing, Burlington, Vermont.

Dodge, Bertha. 1970. *Potatoes and People: The Story of a Plant.* Little, Brown, Boston.

Pringle, C. G. 1871. The breeding of plants by hybridization, selection, etc. *First Annual Report of the Vermont State Board of Agriculture.* J. and J. M. Poland Steam Printing Establishment, Montpelier, Vermont.

———. 1897. My summer in the Valley of Mexico. *Garden and Forest* 10:32–42.

MANUSCRIPTS

A journal of horticultural practice and a record of horticultural science, 1869–1875. Archives, Pringle Herbarium, University of Vermont, Burlington.

Cyrus Pringle's 1899 Diary. Archives, Pringle Herbarium, University of Vermont, Burlington.

Last will and testament of Cyrus Pringle, Pringle Papers, University Archives, University of Vermont, Burlington.

Letter from C. G. Pringle to Asa Gray. September 4, 1884. Pringle Herbarium Archives, University Archives, University of Vermont, Burlington.

Letters from C. G. Pringle to Frank Fenwick Estey. January 19, 1899; February 6, 1899; August 26, 1899; August 7, 1899; August 31, 1899; April 27, 1900; May 13, 1900; June 26, 1900; August 18, 1900; July 27, 1902; November 27, 1905; January 9, 1907. Pringle Papers, University Archives, University of Vermont, Burlington.

Letters from C. G. Pringle to George E. Davenport. March ?, 1884; July ?, 1884; May 12, 1888; May 26, 1895; June 14, 1885; July 28, 1895. Pringle Herbarium Archives, University Archives, University of Vermont, Burlington.
Letter from C. G. Pringle to W. W. Eggleston. June 20, 1902. Pringle Herbarium Archives, University Archives, University of Vermont, Burlington.
Letter from George E. Davenport to C. G. Pringle. December 25, 1883. Pringle Herbarium Archives, University Archives, University of Vermont, Burlington.
Letter from William Stuart to Helen Burns Davis. September 19, 1937. Pringle Herbarium Archives, University Archives, University of Vermont, Burlington.
Record of horticulture, 1867–1868. Pringle Papers, University Archives, University of Vermont, Burlington.

Chapter 8. Windows

Emmons, Ebenezer. 1842. *Geology of New-York, Part III, Survey of the Second Geological District.* White and Visscher, New York.
Ferrin, Clark E. 1873. John Bulkley Perry. *Congregational Quarterly,* April.
Hall, James, 1859. Trilobites of the shales of the Hudson River group. *Twelfth Annual Report of the Regents of the University of New York on the State Cabinet of Natural History*. Albany.
Hitchcock, Edward, Edward Hitchcock, Jr., Albert D. Hager, Charles H. Hitchcock. 1861. *Report on the Geology of Vermont—Descriptive, Theoretical, Economical, and Scenographical.* 2 Volumes. Claremont Manufacturing Co., Claremont, New Hampshire.
Lurie, Edward. 1960. *Louis Agassiz: A Life in Science*. University of Chicago Press, Chicago.
Perry, John B. 1867. Queries on the red sandstone of Vermont and its relations to other rocks. *Proceedings of the Boston Society of Natural History* 11.
———. 1869. A point in the geology of western Vermont. *American Journal of Science* 47:341–349.
———. 1870. A discussion of sundry objections to geology. *Congregational Quarterly,* April.
———. 1871. Natural history of the counties of Chittenden, Lamoille, Franklin and Grand Isle, Vermont. *Vermont Historical Gazetteer* 2:22–28.
Schneer, Cecil J. 1978. The great Taconic controversy. *Isis* 69:173–191

MANUSCRIPTS

Correspondence, 1849–1872. John Bulkley Perry Papers, Special Collections, Bailey-Howe Library, University of Vermont, Burlington.
John Bulkley Perry Papers. Museum of Comparative Zoology, Harvard University, Cambridge, Massachusetts.
Notebook. 1866. John Bulkley Perry Papers. Special Collections, Bailey-Howe Library, University of Vermont, Burlington.
Lectures on geology. John Bulkley Perry Papers, Special Collections, Bailey-Howe Library, University of Vermont, Burlington.
Letter from John Bulkley Perry to George Grenville Benedict. March 2, 1849. John Bulkley Perry Papers, Special Collections, Bailey-Howe Library, University of Vermont, Burlington.
Henry Miles Papers. In possession of Mrs. Forrest Rose, Monkton, Vermont.
Letter from John Bulkley Perry to Prof. Mead of Oberlin College. January 10, 1871. John Bulkley Perry Papers, Special Collections, Bailey-Howe Library, University of Vermont, Burlington.
Manuscript, Hints toward the post-Tertiary history of New England, from personal study of the rocks, with strictures on "The Geology of the New Haven Region," with marginal notes by Louis Agassiz. John Bulkley Perry Papers, Special Collections, Bailey-Howe Library, University of Vermont, Burlington.

U.S. Christian Commission Diary, 1865. John Bulkley Perry Papers, Special Collections, Bailey-Howe Library, University of Vermont, Burlington.

Chapter 9. *Sebamook*

Adams, C. B. 1846. *Second Annual Report on the Geology of Vermont.* Chauncey Goodrich, Burlington, Vermont.
Anonymous. 1906. List of papers now on file. *Vermont Botanical Club Bulletin* 1:17–22.
——. 1951. Annual field meeting, 1951. *Vermont Botanical and Bird Club Bulletin* 19.
——. 1959. *History of Bristol, Vermont.* Outlook Club of Bristol, Bristol, Vermont.
Brainerd, E., L. R. Jones, and W. W. Eggleston. 1900. *Flora of Vermont.* Free Press Association, Burlington, Vermont.
Burns, G. P. 1912. Distribution of bog xerophytes. *Vermont Botanical Club Bulletin* 7:21.
——. 1914. Field work for the botanical club. *Vermont Botanical Club Bulletin* 9:20–22.
Child, Hamilton. 1882. *Gazetteer and Business Directory of Addison County, Vermont for 1881–1882.* Hamilton Child, Syracuse, New York.
Clements, F. E. 1916. *Plant Succession: An Analysis of the Development of Vegetation.* Carnegie Institution of Washington, Washington, D.C.
Colby, Eldridge. 1968. How Bristol Pond became Winona Lake. *Vermont History* 36(3):150–154.
Countryman, W. D. 1978. *Rare and Endangered Vascular Plant Species in Vermont.* U.S. Fish and Wildlife Service in co-operation with the New Hampshire Agricultural Experiment Station, University of New Hampshire, Durham.
Dachnowski, A. P. 1926. Profiles of peat deposits in New England. *Ecology* 7(2):120–135.
Dole, E. J. 1927. The Botanical Club. *Vermont Botanical and Bird Club Bulletin* 12:22.
——. 1937. *The Flora of Vermont.* Free Press Printing, Burlington, Vermont.
Edson, H. A. 1908. Soil reaction in relation to flora. *Vermont Botanical Club Bulletin* 3:34–36.
Gray, Asa. 1877. *Orchis rotundifolia. American Journal of Science* 3(14):72.
Goodale, G. L. 1862. *A Catalogue of the Flowering Plants of Maine.* Portland Society of Natural History, Portland, Maine.
Hills, J. L. and F. M. Hollister. 1912. The peat and muck deposits of Vermont. *Vermont Agricultural Experiment Station Bulletin* 165.
Kirby, W. F., trans. 1907. *Kalevala: The Land of Heroes.* 2 volumes. Everyman Library, New York.
Kittredge, E. M. 1923. Proposed amendment to Act 260 of 1921. *Vermont Botanical and Bird Club Bulletin* 9:12–15.
Longfellow, Henry W. 1879–1883. *The Poetical Works of Henry Wadsworth Longfellow.* Houghton, Osgood, Boston.
Loring, Stephen. 1980. Paleoindian hunters and the Champlain Sea. *Man in the Northeast* 19:15–41.
Miles, Henry. 1875. Is shell marl a fertilizer? *Third Biennial Report of the Vermont State Board of Agriculture.* Charles E. Tuttle, Rutland, Vermont.
Munsill, Harvey M. 1979. *Bristol, Vermont: The Early History.* Bristol Historical Society, Bristol, Vermont.
Perry, J. B. 1868. Indian relics in Swanton. *Proceedings of the Boston Society of Natural History* 15:219–222.
Seymour, F. C. 1982. The flora of New England. *Phytologia Memoirs V.* Plainfield, New Jersey.
Thompson, Zadock. 1842. *Natural history of Vermont.* Charles E. Tuttle, Rutland, Vermont.
Thoreau, H. D. 1909. *The Maine Woods.* T. Y. Crowell, New York.
Tooker, Elisabeth W. 1978. The League of the Iroquois: Its history, politics, and ritual. Pp.

418–441 in *Handbook of North American Indians: Northeast,* Bruce Trigger, ed. Smithsonian Institution, Washington, D.C.
Williams, F. H. 1898. Prehistoric remains of the Tunxis Valley. *American Archaeologist* 2:293–294.

MANUSCRIPTS
Field notebook of D. L. Dutton, 1922–1929. Pringle Herbarium, University of Vermont, Burlington.
Field notebook of W. W. Eggleston. 1892–1903. Pringle Herbarium, University of Vermont, Burlington.
Henry Miles Papers. In possession of Mrs. Forrest Rose, Monkton, Vermont.
John Bulkley Perry Papers. Special Collections, Bailey-Howe Library, University of Vermont, Burlington.
Letter from Asa Gray to C. G. Pringle. July 11, 1880. Pringle Herbarium Archives, University Archives, University of Vermont, Burlington.
Letter from Ezra Brainerd to C. G. Pringle. May 28, 1878. Pringle Herbarium Archives, University Archives, University of Vermont, Burlington.
Letter from Henry Miles to J. B. Perry. December 5, 1868. John Bulkley Perry Papers, Special Collections, Bailey-Howe Library, University of Vermont, Burlington.
Letters from C. G. Pringle to W. W. Eggleston. August 11, 1899; February 24, 1902. Pringle Herbarium Archives, University Archives, University of Vermont, Burlington.
Letters from Charles Sprague Sargent to C. G. Pringle. June 15, 1876; June 21, 1876; July 2, 1876. Pringle Papers, University Archives, University of Vermont, Burlington.

Chapter 10. Blood Streams

Anonymous. 1931. *Rural Vermont: A Program for the Future by 200 Vermonters.* Vermont Commission on Country Life, Burlington, Vermont.
Annual Reports, Eugenics Survey of Vermont, 1927–1931.
Gallagher, Nancy L. 1999. *Breeding Better Vermonters: The Eugenics Project in the Green Mountain State.* University Press of New England, Hanover, New Hampshire.
Hanson, Bertha. 1957. Early Starksboro. *Town of Starksboro Annual Report for 1957.* Starksboro, Vermont.
Ludmerer, K. M. 1972. *Genetics and American Society: A Historical Appraisal.* Johns Hopkins University Press, Baltimore, Maryland.
Perkins, Henry F. 1926. Review of eugenics in Vermont. *Vermont Review* September/October:56–59.
———. 1930. Hereditary factors in rural communities. *Eugenics* 3(8):1–6.
Public Act No. 174—An Act for Human Betterment by Voluntary Sterilization. Public Acts of the general Assembly of the State of Vermont for 1931, pp. 194–196.
Russell, Howard S. 1976. *A Long Deep Furrow: Three Centuries of Farming in New England.* University Press of New England, Hanover, New Hampshire.
Vermont Commission on Country Life Newsletters. 1929–1931.

MANUSCRIPTS
Edith Butler Hannon's manuscript account. Saint Ambrose Parish, Bristol, Conceived in Starksboro. In possession of Ed Hannon, Starksboro, Vermont.

Index